MU120
Open Mathematics

The Open University

GW00640699

Unit 12

Growth and decay

MU120 course units were produced by the following team:

Gaynor Arrowsmith (Course Manager)
Mike Crampin (Author)
Margaret Crowe (Course Manager)
Fergus Daly (Academic Editor)
Judith Daniels (Reader)
Chris Dillon (Author)
Judy Ekins (Chair and Author)
John Fauvel (Academic Editor)
Barrie Galpin (Author and Academic Editor)
Alan Graham (Author and Academic Editor)
Linda Hodgkinson (Author)
Gillian Iossif (Author)
Joyce Johnson (Reader)
Eric Love (Academic Editor)
Kevin McConway (Author)
David Pimm (Author and Academic Editor)
Karen Rex (Author)

Other contributions to the text were made by a number of Open University staff and students and others acting as consultants, developmental testers, critical readers and writers of draft material. The course team are extremely grateful for their time and effort.

The course units were put into production by the following:

Course Materials Production Unit (Faculty of Mathematics and Computing)

Martin Brazier (Graphic Designer)
Hannah Brunt (Graphic Designer)
Alison Cadle (TEXOpS Manager)
Jenny Chalmers (Publishing Editor)
Sue Dobson (Graphic Artist)
Roger Lowry (Publishing Editor)
Diane Mole (Graphic Designer)
Kate Richenburg (Publishing Editor)
John A.Taylor (Graphic Artist)
Howie Twiner (Graphic Artist)
Nazlin Vohra (Graphic Designer)
Steve Rycroft (Publishing Editor)

This publication forms part of an Open University course. Details of this and other Open University courses can be obtained from the Student Registration and Enquiry Service, The Open University, PO Box 197, Milton Keynes, MK7 6BJ, United Kingdom: tel. +44 (0)870 333 4340, e-mail general-enquiries@open.ac.uk

Alternatively, you may visit the Open University website at http://www.open.ac.uk where you can learn more about the wide range of courses and packs offered at all levels by The Open University.

To purchase a selection of Open University course materials, visit the webshop at www.ouw.co.uk, or contact Open University Worldwide, Michael Young Building, Walton Hall, Milton Keynes, MK7 6AA, United Kingdom, for a brochure: tel. +44 (0)1908 858785, fax +44 (0)1908 858787, e-mail ouwenq@open.ac.uk

The Open University, Walton Hall, Milton Keynes, MK7 6AA.

First published 1996. Second edition 2003. Reprinted with corrections 2004, 2005, 2006.

Edited, designed and typeset by The Open University, using the Open University TEX System.

Printed and bound in the United Kingdom by The Charlesworth Group, Wakefield.

ISBN 0 7492 5777 6

2.4

Contents

Study guide

There are five sections in this unit. The study plan below shows details of approximate timings.

None of the sections makes use of audio or video bands, though the TV programme 'The True Geometry of Nature' is relevant to Section 1.2.

The calculator is required at many points in the unit; in particular, Section 3 and part of Section 5 consist almost entirely of activities based on the calculator. The Course Reader is used in Section 4.

There is only one Activity Sheet associated with this unit. It may be useful for recording definitions and/or explanations of new terms, so as you work through the unit, make appropriate entries.

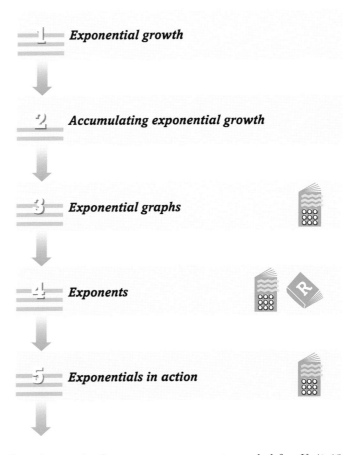

1 *Exponential growth*

2 *Accumulating exponential growth*

3 *Exponential graphs*

4 *Exponents*

5 *Exponentials in action*

Summary of sections and other course components needed for *Unit 12*.

Introduction

Back in Chapter 1 of the *Calculator Book*, you carried out some investigations into the use of scientific notation. One of these involved an extremely generous monarch, Queen Calcula of Sumwhere, who was prepared to offer you one gold piece on the first day of the month, two on the second, four on the third, eight on the fourth, and so on—ending this generosity on the last day of the month. In the course of the investigation, you had to find out how many gold pieces were on offer each day. It turned out that on the last day of the month (assuming it to be 31 days long), there were $1\,073\,741\,824$, or about 1.07×10^9, gold pieces available. In another investigation, you were asked to calculate how many ancestors you have, going back to the thirtieth generation.

Both of these investigations involved a number (of gold pieces or of ancestors) that increased step by step, doubling each time: the number of gold pieces offered by Queen Calcula on any day of the month was *twice* the number that she had offered on the previous day; the number of your great-grandparents was *twice* the number of your grandparents.

These are examples of change according to a pattern called *exponential growth*. The term 'exponential' applies to any process in which something changes according to the rule that the quantity at any stage is a fixed positive multiple of the quantity at the previous stage.

It so happens that in both of the above examples the factor by which the quantity grows at each stage (the multiplier) is 2. However, the word 'exponential' is used whatever the factor is. For instance, in *Unit 9*, you saw that the frequency of a note sounding one semitone higher than the previous note in the Western chromatic scale is always $\sqrt[12]{2}$ times the previous frequency: this is an example of exponential growth with a growth factor of $\sqrt[12]{2}$, or 1.059 (to 3 decimal places).

Exponential change can sometimes involve decay rather than growth. In that case, the quantity concerned will *decrease* step by step. The fixed factor will then be a number such that the quantity at a given stage is a *fraction* of that at the preceding stage; this means that the factor must lie between 0 and 1. As you will see in Section 5, carbon dating works on the principle that the amount of radioactivity in carbon decays exponentially, with a multiplicative factor of 0.999 879 each year.

The main aims of this unit are to describe and explain the special features of exponential change, and to give examples of situations in which exponential change occurs.

Mathematical modelling was introduced in *Unit 10*.

When a practical situation is described as 'exponential', this simply refers to the way in which something grows or decays. A mathematical description of the situation then leads to an exponential model. However, as with any mathematical model, a careful analysis of the observed changes may reveal that what is going on is actually more complex than the model suggests.

The key features of an exponential model are that:
- the situation can be modelled by a sequence of stages (steps, generations, ...);
- moving from one stage to the next involves multiplying by a fixed positive factor, called the *growth factor*;
- in the case of exponential *growth*, the growth factor is greater than 1, whereas in exponential *decay*, the factor is between 0 and 1.

You will see later in the unit that for some exponential models, rather than there being a sequence of distinct steps, the exponential growth or decay occurs continuously. But in the early part of this unit, only situations involving distinct stages, steps or generations will be considered.

1 Exponential growth

Aims   This section presents a number of examples of exponential growth and decay, and aims to show how to model this kind of change mathematically.   ◇

1.1 Examples of exponential change

Examples of exponential change are widespread and include, among other things, the distribution of chain letters, the calculation of interest on savings or loans, and the decay of radioactive atoms. All show the same pattern of repeated multiplication by a constant factor—a pattern that is characteristic of exponential growth or decay.

Chain letters

A chain letter typically asks its recipients to send a copy of the same letter to a certain number of people: five people, perhaps. These new recipients are each required to send further copies on to five of their friends. So, at each stage, five times as many people will receive a letter as received one at the previous stage. The number of letters at each stage increases exponentially (with a growth factor of 5 in this case) so long as all of the recipients obey the rules.

The same principle applies to 'pyramid selling', where the base of the pyramid grows as more and more people are drawn in.

Greenfly

Why do roses get covered with greenfly so quickly? Because greenfly breed very fast: they can reproduce asexually, and each individual in each new generation is able to reproduce when it is just 24 hours old. Each individual produces (roughly) the same number of young, so the population of greenfly on a rose bush grows exponentially—until the gardener, or the ladybirds, do something about it.

Trees

In the autumn, buds develop at and near the end of each twig on a deciduous tree. Next spring (provided the tree survives), each of these buds sprouts into a new twig. One obvious question is: how does the number of twigs or, equivalently, buds on a growing tree change with its age? This is one of the features that governs the overall shape of a tree; (another feature is the geometry of the twigs—their lengths and the angles between them).

An idealized tree is not subject to the vagaries of disease or the weather.

Consider an idealized tree with a single bud conveniently located at the end of each twig. In the spring, each of these buds sprouts a new twig, while the old twig continues to grow, forming a fork (see Figure 1). Thus the number of twigs doubles each year (as do the number of buds), and hence the growth is governed by the fixed multiplicative factor 2. So, whether you are counting twigs or buds, the growth of this idealized tree is exponential in form.

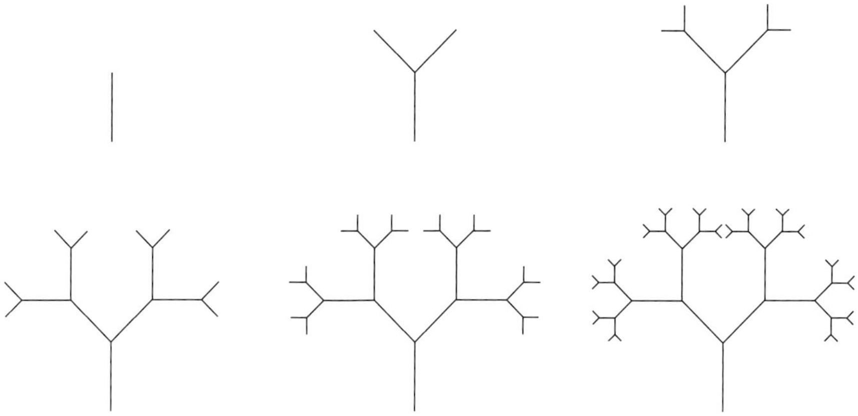

Figure 1   Idealized trees exhibiting exponential growth.

Interest

A sum of money in a typical savings account, attracting compound interest at a fixed rate, grows exponentially: interest earned is added to the sum in the account at the end of a fixed period (often a year), and then in the subsequent period, interest is calculated on the basis of the new total, and so on. This is an important example of exponential growth and is covered in some detail in Example 2, later in this section.

Radioactive decay

The atoms of some elements, such as uranium, occur in a number of different forms called *isotopes*. Each particular isotope is distinguished by the composition of the atom's nucleus. The nucleus of an isotope may contain too many sub-particles (such as neutrons) to cohere together satisfactorily; the isotope is then said to be unstable, and may spontaneously convert into something else, releasing alpha (α) or beta (β) particles and, perhaps, gamma (γ) rays. An isotope that behaves in this way is said to be *radioactive*, and the process by which a radioactive isotope breaks down is called *radioactive decay*.

When physicists first began investigating the structure of the nucleus, they discovered several new particles or rays and named each with a Greek letter.

Imagine a collection of atoms of a radioactive isotope of some element. A constant proportion of the atoms decay in any given time interval. The number of radioactive atoms will, therefore, decrease over time as more and more of them break down. It has been found, both theoretically and by empirical observation at regular intervals, that radioactive isotopes decay exponentially. In other words, the number of atoms is regularly and repeatedly multiplied by some constant factor that lies between 0 and 1.

The Sierpinski carpet

Begin with a black square; subdivide it into nine smaller squares and paint the middle one white; subdivide each of the remaining eight black small squares into nine yet smaller squares and paint the middle one white; continue in this way as shown in Figure 2. As the process is repeated, the black area is repeatedly multiplied by 8/9, and so it decreases exponentially.

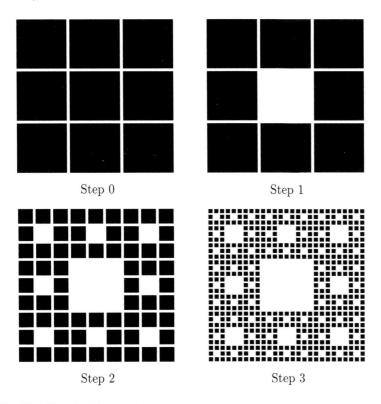

Step 0	Step 1
Step 2	Step 3

You can find a program to create a Sierpinski triangle in the Activities section of the calculator manual.

Figure 2 The Sierpinski carpet.

The examples of exponential change that have been outlined all involve the idea that, in going from one stage to the next, a quantity is multiplied by a fixed growth factor. When describing biological examples such as greenfly populations, the term 'generation' is often used, rather than 'stage'. However, the words 'population' and 'generation' are also used metaphorically in various other exponential situations—the so-called population might consist of euros, sticky-buds, letters or atoms, as well as people, microbes or greenfly.

In the examples of exponential change considered in this unit, the *number* of things undergoing change will usually be referred to as the *size of the population*. For most of the unit, each population will be modelled as changing in size in a discrete, step-by-step manner; each step will usually be referred to as a *generation*. In this terminology, the defining characteristic of such *exponential* change is that the size of the population in one generation is a fixed positive multiple of the size of the population in the previous generation.

Note that this unit often uses the general term 'exponential *growth*' despite the fact that the exponential change involved may be decay rather than growth; this is done to save repeatedly having to write 'growth or decay'.

At this juncture, it is worth summarizing some key points about exponential change:

A similar example in mathematics is the use of the term 'acceleration' to cover both speeding up and slowing down.

- If the population is increasing, there is positive growth; if it is decreasing, there is negative growth (or decay).
- The factor by which the population size is multiplied in every generation is called the exponential *growth factor*.
- If the growth factor is greater than 1, then the population grows.
- If the growth factor lies between 0 and 1, then there is negative growth—the population decays, decreasing at each stage.
- If the growth factor is 1, the population remains unchanged—there is no exponential change.
- A growth factor of 0 will reduce the population to 0 in a single generation; this would not normally be referred to as exponential change.
- A growth factor of less than 0 would make little sense in a practical situation; it would result in a population size that was alternately positive and negative in successive generations.

Activity 1 *Is the growth exponential?*

Here are some examples of situations, taken from earlier units, which involve change. Which (if any) of them are examples of exponential growth? For those that are, identify the population and the growth factor.

(a) A culture of bacterial cells in a test tube, where the height of the top of the cell culture changes from 1 mm to 2 mm to 4 mm to 8 mm in successive minutes.

(b) An object falling from a height and accelerating downwards with the acceleration due to gravity, so that it falls further in every successive second.

(c) 'As I was going to St Ives,
   I met a man with seven wives;
   each wife had seven sacks,
   each sack had seven cats,
   each cat had seven kits;
   kits, cats, sacks, wives:
   how many were going to St Ives?'

   The change involved here is in the number of things referred to when moving from line to line of the riddle.

(d) The world record time for running the mile, which decreased from about 4.2 minutes in 1920, to 4 minutes in 1950, to 3.8 minutes in 1980.

Activity 2   *Find some yourself*

Find some examples of exponential growth for yourself. If you keep an eye on the newspapers and TV, you will probably spot some, because the media are quite fond of the word 'exponential' (for example, 'the amount of computing power available per thousand pounds is growing exponentially'). Also consider whether the word is always used correctly.

1.2   *Formulas for exponential growth*

How do you calculate how many ancestors you have in each previous generation? Well, in the first generation back, you have two ancestors (your biological parents); in the second, you have four (your grandparents, or your parents' parents); in the third, there are eight (your great-grandparents, or your grandparents' parents); and so on.

This assumes that no-one appears more than once in the family tree.

Therefore, the pattern goes:

1 generation back		2 ancestors,
2 generations back	$2 \times 2 = 2^2 =$	4 ancestors,
3 generations back	$2 \times 2 \times 2 = 2^3 =$	8 ancestors,
4 generations back	$2 \times 2 \times 2 \times 2 = 2^4 =$	16 ancestors,

and so on.

This can be expressed by using a general formula that contains a variable n, which stands for the generation being described. Thus the number of ancestors in the nth generation back is 2^n. Substitute in the first few values of n to check that the formula does, in fact, give the values shown above.

Activity 3   *Chain letters*

Consider a chain letter that contains the instruction 'send a copy of this letter to five of your friends'. To keep things simple, assume that all the participants obey the instruction and that no-one is sent the letter by more than one other person in the chain. The person who sets the chain letter in motion at the first stage sends out five letters; each recipient then sends out five letters, and so on.

How many letters are sent at the third stage? How many at the seventeenth stage? How many at the nth stage?

The findings in the two examples considered above can be generalized. Suppose the population size (of ancestors, letters, or whatever) in a particular generation or stage is represented by P, and the number of that generation or stage is represented by n. Then, for the problem of the number of ancestors, you can write

$$P = 2^n.$$

So if you want to calculate the number of ancestors you have ten generations back, then you just substitute 10 for n in this formula, to get $P = 2^{10} = 1024$.

In a similar fashion, the formula for the chain letter problem is

$$P = 5^n,$$

where P is the population or number of letters sent at the nth stage.

Activity 4 *Chain mail*

How many chain letters are sent at the tenth stage?

As already mentioned, descriptions of exponential growth are often couched in terms of the words 'population' and 'generation' even in cases as diverse as the number of twigs on a tree or the number of chain letters sent. With 'population' and 'generation' used in this general sense, the size of the population in any case of exponential growth depends on which generation you look at; in other (mathematical) words, the size of the population is a function of the generation. When it is desirable to emphasize this dependence, the notation is extended a little, and $P(n)$ is written for the population size, rather than just P. This can be read as 'P of n', or as 'the population size in generation n'. This extended notation helps to make explicit which generation is being talked about.

Sometimes the notation P_n is used instead of $P(n)$.

You can take advantage of this notation to create a useful shorthand for a phrase like 'the number of ancestors you have in the tenth generation back': this would be simply $P(10)$. Hence the general formula for the population size of ancestors can be written as

$$P(n) = 2^n.$$

The word 'exponent' comes from the Latin verb pono, meaning 'I place', with the prefix ex- meaning 'out'. It is an exponent because you place it out (of the way).

The variable n in an expression like 2^n or 5^n is referred to as an *exponent*, and the term *exponential* simply means 'involving an exponent'. A function in which the independent variable appears as an exponent is called an *exponential function*. So, for instance, $y = 5^x$ is an exponential function, but $y = x^5$ is not.

In this unit you will be dealing a lot with exponentials like 5^x, 10^x, b^x and so on; that is, with expressions in which one value is raised to the power of another. The terms *base* and *exponent*, or *power*, are used to distinguish the components of such expressions. Thus, in b^x, b is the base and x is the exponent or the power, as summarized at the top of the page opposite. Note that the terms 'exponent' and 'power' will be used interchangeably throughout the rest of this unit.

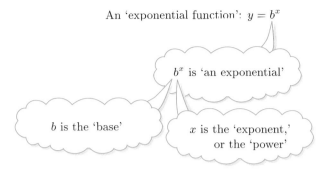

You have now met several examples of exponential growth, so you should be beginning to get a clearer idea about it. When learning about a new concept, it often helps to know what it is *not*, as well as what it is. The next two examples allow you to compare non-exponential and exponential growth. The topic is a financial one: interest earned by a savings account or a similar investment.

Example 1   *Savings account: simple interest*

Consider an investment where the interest accrued is paid out to the investor at the end of each year. In this situation, the account earns *simple interest* — the interest earned is regarded separately from the deposit and does not itself gain interest. Suppose that this savings account offers a fixed simple interest rate of 10% annually. If a sum of £200 is deposited initially, how does the investor's money grow?

Solution

At the end of one year, the interest is

$$£200 \times \tfrac{10}{100} = £200 \times 0.1 = £20,$$

and the investor now has £220. At the end of the second year, an *identical* amount of interest is earned, so the investor's money increases to £200 + £20 × 2 = £240. This is repeated at each year-end.

If $P(n)$ represents the total amount of the investor's money at the end of the nth year, then a formula for calculating $P(n)$ is

$$P(n) \quad = \quad £200 \quad + \quad £20 \times n$$

$P(n)$	£200	£20 × n
↑	↑	↑
the amount after n years	the initial deposit	£20 for each year

As you can see, this is not exponential growth; it is, in fact, an example of linear growth, as you should recognize from *Unit 10*.

Now here is a contrasting case.

Example 2 *Savings account: compound interest*

Most savings accounts pay *compound interest*. This means that once the interest is earned, it is added to the sum already in the account, and then this new *total* sum earns interest. Consider the case discussed in the previous example, but suppose that the account now pays compound rather than simple interest. This time, at the end of each year, the interest is *added* to the balance of the account. How do the investor's savings grow this time?

Solution

At the end of the first year, the balance of the account increases by 10%, to give 110% of the balance at the start. This corresponds to an increase by a factor of $\frac{110}{100}$, or 1.1, resulting in a balance of £200 × 1.1 = £220.

At the end of the second year, the balance is multiplied by a further factor of 1.1, giving

$$£200 \times 1.1^2 = £242.$$

At the end of the third year, the balance is

$$£200 \times 1.1^3.$$

At the end of the nth year, the balance is

$$£200 \times 1.1^n.$$

If $P(n)$ represents the amount of money in the account after n years, then:

$$P(1), \text{ amount of money after 1 year } = £200 \times 1.1,$$
$$P(2), \text{ amount of money after 2 years} = £200 \times 1.1^2,$$
$$P(n), \text{ amount of money after } n \text{ years} = £200 \times 1.1^n.$$

This satisfies the condition for exponential growth in that the money grows each year by a fixed growth factor, 1.1.

Note that Example 2 differs from the examples you have met already by the presence of the coefficient 200, which is the sum initially invested. You might find it helpful to think of the term 1.1^n as representing the amount to which each £1 initially invested would have grown. So the £200 will grow to 200 times this amount.

A general point to remember here is that compound interest results in exponential growth, whereas simple interest results in linear growth.

Activity 5 More compound interest

Suppose that £500 is invested at 4% compounded annually over 5 years. What is the balance of the account at the end of this period?

What would the balance have been at 4% simple interest?

Activity 6 More on money

You put £A into a savings account that gives 3% compound interest annually. How much will there be in the account at the end of the nth year after you deposited the money?

Activity 7 It's the rich ...

A successful rock star wishes to give her newly born son a gift of £1 000 000 when he comes of age in 18 years' time. She finds an account that guarantees to pay 5% per year, compounded annually, for the whole of this period. She wants to know how much she must invest now in order to achieve her target.

If the amount she invests is £A, write down an expression for what this will be worth in 18 years' time. Set this expression equal to £1 000 000, and so find the value of A.

Activity 8 Simplified greenfly

The simplified greenfly is a mythical insect—but designed to help you understand how real populations (of insects, birds, yeast cells, crocodiles and even, possibly, people) grow. The greenfly breeds asexually. Each adult simplified greenfly has three offspring and then dies. The offspring reach maturity on the day after their birth, and each has three offspring, and then itself dies that day; and so on. On 1 May, I find that there are 17 newborn simplified greenfly on my favourite rose bush. Write down a formula for the number of simplified greenfly there would be on the bush n days later if I do nothing about it.

Use the formula to predict approximately how many greenfly there will be on the bush on 11 May.

Here is another situation that leads to exponential growth.

Example 3   *Mathematical snowflakes*

The snowflake curve is an example of a *fractal* which is discussed in the TV programme *The True Geometry of Nature*.

There is a rather beautiful geometric object called a snowflake curve, which can be constructed step-by-step as shown in Figure 3. Start with an equilateral triangle (a triangle whose sides are all the same length). Divide each side into three equal parts. On the middle third of each side, draw another equilateral triangle pointing outwards (each of the sides of this triangle will be one third of the length of the original one), and then erase the middle third itself. The result is a six-pointed star. Next take each side (that is, each straight-line section) of the star, divide it into three equal parts, and replace the middle third as before. You get a shape that has no name, but is beginning to look like a snowflake.

Figure 3   Constructing a snowflake curve.

Now repeat the whole process many times. The ultimate curve that you would get if you subdivided the sides indefinitely is called a snowflake curve. It may seem a bit odd to call this figure a 'curve', since it is made up of straight lines, but after the process has gone through many stages the jagged nature of the shape takes on a smoother appearance that is more characteristic of a curve.

The snowflake curve is said to be 'self-similar': if you take any little bit of the curve and magnify it suitably, what you get is essentially indistinguishable from a large bit of the curve. The snowflake curve is rather like a cauliflower in this respect. It is also like the leaves of some plants, where the leaves have toothed edges, and on close inspection the teeth themselves are toothed, and so on.

One interesting question about a snowflake curve is: how long is it? How does the perimeter of a snowflake curve—the total length of its sides—change, stage by stage?

Solution

At each stage of the construction, some straight lines are replaced by another set of lines with kinks in them. The general replacement step is shown in Figure 4.

Figure 4   Constructing a snowflake curve: replacing a line.

For each straight line, the middle third is replaced by something twice as long. Thus the three-thirds length of the original line turns into four-thirds as a result of this replacement. In other words, each line in the figure is replaced by something that is four-thirds as long. This means that the total length of the perimeter of the figure is multiplied by the factor $\frac{4}{3}$ at each stage.

Suppose, for example, that the original equilateral triangle has sides of length 1 unit, so its perimeter is 3 units. Then the perimeters of the successive figures in the construction of the snowflake curve are

$$3 \times \tfrac{4}{3}, \qquad 3 \times \left(\tfrac{4}{3}\right)^2, \qquad 3 \times \left(\tfrac{4}{3}\right)^3, \qquad 3 \times \left(\tfrac{4}{3}\right)^4, \qquad \ldots .$$

As you can see, the perimeter grows exponentially, with a growth factor of $\frac{4}{3}$. The coefficient of 3 in these expressions is simply the perimeter of the original triangle.

Activity 9 *More on the snowflake curve*

(a) If $P(n)$ is the perimeter of the figure created in the nth stage of generating a snowflake curve, write down a formula for $P(n)$.

(b) Suppose that the sides of the original triangle were each $\frac{1}{3}$ of a unit in length. What would be the perimeter of the figure at the nth stage?

Suppose that the perimeter of the original triangle were 5 units. What would be the perimeter at the nth stage?

(c) Write down a formula that gives the perimeter of the figure obtained in the nth stage if the perimeter of the original triangle were l.

You have now investigated five examples of exponential growth—chain letters, number of ancestors, savings accounts (based on compound interest), greenfly populations and snowflake curves. In each case, there is a formula that defines the population size, $P(n)$, at stage n, where $n = 1$ corresponds to the first step in the growth or change process:

- *Number of ancestors*
  The formula for the number of ancestors n generations back is
  $$P(n) = 2^n.$$

- *Chain letters*
  The formula for the number of letters sent at stage n, assuming that each person sends on the letter to five new people, is
  $$P(n) = 5^n.$$

- *Savings accounts (based on compound interest)*
  The formula for the amount of money in a savings account after n years, when the account starts off with a deposit of £200 and earns compound interest at 10% per annum is

  $$P(n) = \pounds200 \times 1.1^n.$$

- *Greenfly population*
  The formula for the number of greenfly after n days, starting off with 17 greenfly and replacing each greenfly by three new offspring in each generation, is

  $$P(n) = 17 \times 3^n.$$

- *Snowflake curve*
  The formula for the perimeter, at stage n, of a snowflake curve that starts off with a perimeter of l and increases by a growth factor of $\frac{4}{3}$ is

  $$P(n) = l \times \left(\tfrac{4}{3}\right)^n.$$

From these formulas, it is possible to deduce some useful general formulas that hold for any exponential model.

Assume that

$$
\begin{aligned}
n &= \text{number of generations or stages that have occurred,} \\
P(n) &= \text{population size in generation } n, \\
a &= \text{initial population size,} \\
b &= \text{growth factor.}
\end{aligned}
$$

It follows from the five examples summarized above that the general exponential model can be described by the formula

$$P(n) = ab^n. \tag{1}$$

Now consider what happens initially; that is when $n = 0$, which corresponds to generation 0. The general formula (1) then gives the initial population as

$$P(0) = a \times b^0.$$

Since the initial population, $P(0)$, is a, it follows that whatever the value of b, $b^0 = 1$.

An explanation of why something raised to the power 0 equals 1 is given in Section 4, but you might like to check this on your calculator now.

If a is replaced by $P(0)$ in formula (1), then

$$P(n) = P(0)b^n. \tag{2}$$

Expressing the exponential model as shown in formula (2) has the advantage of emphasizing that the value of a is actually the population size in generation 0.

Another useful connection is that the population size in generation $(n + 1)$, that is $P(n + 1)$, results from applying the growth factor, b, to the population size of the previous generation, $P(n)$. So

$$P(n + 1) = b \times P(n),$$

or

$$P(n + 1) = bP(n).\tag{3}$$

The three numbered formulas provide useful alternative ways of describing the general exponential model. Some of these formulas may be more applicable to certain models than others. For example, in modelling the daily pay out by Queen Calcula, described in the Introduction, formula (3) is most useful. On the other hand, formula (2) is not suitable because $n = 0$ is not included in the Queen Calcula model (remember that for a month of 31 days, n runs from 1 to 31, and 0 is not included).

1.3 *Collared doves: an exponential model*

This section involves an example of the use of an exponential model to describe the growth of a population of living things—in this case, birds. The text in the box overleaf is taken from the book *Models in Biology* by D. Brown and P. Rothery (Wiley, 1993). (A few minor changes have been made to bring the terminology and notation into line with this unit.)

When dealing with processes that occur in nature, modelling becomes more complicated than it is for situations under controlled conditions in the mathematical laboratory. One very useful technique is illustrated in the passage in the box: the use of what the authors call a 'word model'. The authors express in words the basic relationship between the number of breeding females in one generation and the number in the previous generation, before they attempt to formulate the relationship algebraically. This should help you to understand the derivation of the formula $P(n + 1) = \left(k + \frac{1}{2}ml\right) P(n)$, which is the key relationship in Brown and Rothery's mathematical model. This method is similar to that used in *Unit 8* where there was a transition from a word formula for a 'think of a number' sequence to an algebraic expression. It is a good technique to use when you are formulating a mathematical model yourself.

Compare this formula with formula (3) above.

After the box, there is an activity to test your understanding of the modelling process. It should also suggest to you things to look out for when reading similar accounts in the future.

THIRTY-THREE THOUSAND, FIVE HUNDRED AND SIXTY-TWO...

Collared doves

Following a dramatic expansion of its range across Europe from the Balkans to the North Sea in under 20 years, the collared dove, *Stretopelia decaocto*, bred for the first time in England (in Norfolk) in 1955. The subsequent increase in numbers over the next ten years is summarized in Figure 5.

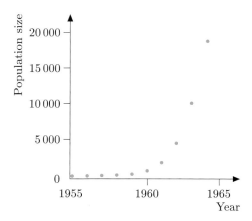

Figure 5   Number of collared doves in the British Isles: the population size is the number of adults at the beginning of the breeding season.

By 1964, the species had reached most areas of Britain and the population size, measured as the number of adults at the beginning of the annual breeding season, had increased from four to an estimated 18 855. To explain the rapid increase in population size, a simple model will be built which incorporates information on survival and reproduction. Each year, a number of breeding pairs rear young which are capable of breeding as adults in their first year after fledging (learning to fly). The number of adult females in the breeding population in a particular year is made up of the surviving adult females from the previous year plus the number of their female offspring that survive. A word description, or word model, for the female population change is

$$
\begin{pmatrix} \text{number of} \\ \text{adult females} \\ \text{in breeding} \\ \text{population} \\ \text{in year } (n+1) \end{pmatrix} = \begin{pmatrix} \text{number of} \\ \text{adult females} \\ \text{in breeding} \\ \text{population} \\ \text{in year } n \end{pmatrix} \times \begin{pmatrix} \text{proportion of} \\ \text{adults} \\ \text{surviving to} \\ \text{following} \\ \text{year} \end{pmatrix}
$$

$$
+ \begin{pmatrix} \text{number of} \\ \text{young females} \\ \text{produced in} \\ \text{year } n \end{pmatrix} \times \begin{pmatrix} \text{proportion of} \\ \text{young} \\ \text{surviving to} \\ \text{following} \\ \text{year} \end{pmatrix}.
$$

Developing this model for the population change involves considering the simple case in which survival rates and birth rates do not vary from year to year. A constant proportion, k, of adults breeding in one year survive to breed in the next year, and a constant proportion, l, of young survive to become adults and breed at the end of their first year. Each female produces a constant number, m, of young, of which half are assumed to be females.

The word model can now be written as an equation for the number of adult females, $P(n+1)$, in the breeding population in year $(n+1)$:

$$P(n+1) = P(n) \times k + P(n) \times \tfrac{1}{2}m \times l$$
$$= kP(n) + \tfrac{1}{2}mlP(n),$$

or

$$P(n+1) = \left(k + \tfrac{1}{2}ml\right)P(n).$$

This step involves taking out the common factor $P(n)$.

This formula reduces to a model of exponential growth of the form

$$P(n+1) = bP(n),$$

Recall formula (3).

where the factor b is related to the survival and birth rates by the relation

$$b = k + \tfrac{1}{2}ml.$$

Hofstetter studied collared doves in Germany in 1954 and estimated average survival rates of 86% for adults and 60% for juveniles in their first year. Collared doves lay clutches of two eggs, although most pairs lay two clutches and some lay more, so, on average, each pair rears four young. For a population in which $k = 0.86$, $l = 0.60$ and $m = 4$, the calculated exponential growth factor, b, is 2.06 per annum.

Now work through Activity 10, which explores various aspects of the model described in the box.

Activity 10 *Investigating the model*

(a) What does the term $kP(n)$ correspond to in the word model?

Explain how the term (number of young females produced in year n) × (proportion of young surviving to following year) is translated into symbols to complete the algebraic version of the model.

(b) On the basis of observations of collared doves, the values of k, l and m in the population model were estimated to be 0.86, 0.60 and 4, respectively. Using these estimates, how is the value of 2.06 obtained for the growth factor in the population model?

(c) From Figure 5, the population size in 1960 is roughly 1000. Compare the model's predicted results with the measured population sizes for the years 1961 to 1964, taking 1000 as the initial value of the population in 1960.

(d) The population size is defined as 'the number of adults at the *beginning* of the annual breeding season'. Why is it necessary to be so careful about the definition of the population size? How does the population size vary *during* the year, as opposed to from one year to the next?

(e) Can you suggest any reasons why the population of collared doves in the British Isles grew so rapidly once the doves had become established? Do you think that the population could continue to grow exponentially?

In part (c) of Activity 10 the model takes the initial population as that in 1960, but for other applications it may be appropriate to start at another date, say, 1955—in that case the population in 1960 would correspond to $n = 5$. The choice of an origin depends on the purpose of the model, and a certain amount of common sense is needed in selecting the starting point.

In summary, this section has described exponential growth for real world situations and has expressed this growth in terms of mathematical relationships. The characteristic feature of exponential growth is that the population size from one generation to the next changes by a fixed multiple, the growth factor. This relationship can be expressed as

$$P(n + 1) = bP(n),$$

where b is the growth factor.

An explicit formula for the population size, $P(n)$, in generation n can be stated in terms of the growth factor, b, and the initial population size, a, as

$$P(n) = ab^n.$$

However, since a is $P(0)$, the formula can also be written as

$$P(n) = P(0)b^n.$$

Outcomes

After studying this section, you should be able to:

◇ explain what is meant by 'exponential growth', and recognize examples of such growth (Activities 1, 2 and 5);

◇ write down and use general formulas for the size of a population that is growing exponentially (Activities 3–9);

◇ interpret exponential models (Activity 10).

2 Accumulating exponential growth

Aims   This section introduces the idea of adding together the terms of an exponential sequence.   ◇

Think back to the bountiful Queen Calcula described in the Introduction. As you may recall, the queen's generosity fitted the pattern of exponential change with a growth factor of 2. Thus the queen was prepared to give you one gold piece on the first day of the month, two on the second day, four (or 2^2) on the third day, eight (or 2^3) on the fourth day, and so on. A clear pattern is emerging here, indicating that on the nth day of the month you will receive 2^{n-1} gold pieces (because the power to which you raise 2 in this case is one less than the day number).

However, this does not tell you how many gold pieces you would have received *in total* by a given day. The total that you would have received by the nth day is the cumulative sum of the amounts received on days 1 to n. This cumulative sum is denoted by $S(n)$ and is given by

$$S(n) = 1 + 2 + 2^2 + \cdots + 2^{n-1}.$$

Is it possible to produce a simpler formula for this sum?

When tackling an investigation of this sort, an effective strategy is to:

- try out the formula with simple numbers;
- present the results clearly and systematically in a table;
- look for a pattern.

Table 1 shows the results for the first few values of n.

Table 1

Day, n	Number of gold pieces received	Total number of gold pieces received, $S(n)$
1	$2^0 = 1$	1
2	$2^1 = 2$	$1 + 2 = 3$
3	$2^2 = 4$	$1 + 2 + 4 = 7$
4	$2^3 = 8$	$1 + 2 + 4 + 8 = 15$
5	$2^4 = 16$	$1 + 2 + 4 + 8 + 16 = 31$

Look closely at the final column of this table. Notice that the numbers there (1, 3, 7, and so on) are just one short of being powers of 2. To take a particular example, the sum for the fifth day, $S(5) = 2^5 - 1 = 32 - 1 = 31$. Hence the total on day n appears to be

$$S(n) = 2^n - 1. \qquad (4)$$

Although this formula works for specific numbers, a more general proof is needed to show that it is always true for Queen Calcula and, indeed, for similar exponential situations where the growth factor is 2. Such a proof is set out below.

The proof consists of the following three stages:

(i)   Write down two formulas for $S(n+1)$, each involving $S(n)$.

(ii)   Create an equation for $S(n)$ by setting the two formulas equal to one another.

(iii) Solve the equation for $S(n)$.

This is a useful approach and one that you are likely to meet again.

You can use this approach to prove that formula (4) for the cumulative sum always holds in cases where the growth factor is 2 and the initial population size is 1. The stages involved in the proof are as follows:

(i)   You know that tomorrow's total, $S(n+1)$, is today's total, $S(n)$, plus tomorrow's amount, 2^n. Thus

$$S(n+1) = S(n) + 2^n.$$

Another way of looking at the relationship between $S(n)$ and $S(n+1)$ is to write $S(n+1)$ as

$$\begin{aligned} S(n+1) &= 1 + 2^1 + 2^2 + \cdots + 2^{n-1} + 2^n \\ &= 1 + 2(1 + 2 + 2^2 + \cdots + 2^{n-1}) \\ &= 1 + 2S(n). \end{aligned}$$

Think carefully about why this step is true.

(ii)   Setting the right-hand sides of these two formulas for $S(n+1)$ equal to one another produces an equation involving only $S(n)$:

$$1 + 2S(n) = S(n) + 2^n.$$

(iii) Solve for $S(n)$:

$$\begin{aligned} 1 + S(n) &= 2^n & \text{(by taking } S(n) \text{ from both sides)} \\ S(n) &= 2^n - 1 & \text{(by taking 1 from both sides).} \end{aligned}$$

This constitutes a formal proof that $S(n) = 2^n - 1$ holds for all positive integer values of n when the growth factor is 2 and the initial population size is 1. Shortly, more general formulas for $S(n)$ will be developed that hold for any growth factor and any initial population size.

Activity 11   *February—a golden month*

How many pieces of gold in total will Queen Calcula hand over in the month of February?

Before going on to establish more general formulas for $S(n)$, it is worth giving some thought to the question of notation in order to avoid any possible confusion.

A note on differing conventions

There is no single 'correct' notation for modelling growth. In particular, you will find that there appears to be inconsistency about the meaning attached to the stage (or year or generation) number, n. On some occasions it is convenient to think of the starting population as occurring in stage (or year or generation) 0, so here the model starts at $n = 0$. In other situations, the first population value being considered is referred to as the population where $n = 1$ (thus the first gold piece in the Queen Calcula example corresponded to day 1, not day 0), so here the model starts at $n = 1$. Clearly, this can be confusing. However, because the contexts of these problems vary, such ambiguity is hard to avoid. It is important that you check the small print of the examples very carefully and be clear in your mind whether the model starts at $n = 0$ or $n = 1$. In this section, most of the examples start from $n = 1$.

Just as in the tale of the queen's pieces of gold, there will be other situations where you want to know the cumulative total of the population sizes when a population is growing exponentially. For example, in the case of the chain letters, it is perhaps more interesting to know the total number of people involved in the chain, rather than the number of letters posted in each generation. Therefore, it is useful to have a general formula that can serve to calculate the cumulative sum in a variety of situations. So, consider a general case where a population that is growing exponentially has a growth factor b. What is the cumulative sum, $S(n)$?

Start by taking the simple case where the size of the population at the outset is 1 when $n = 1$. Then the population sizes will be 1, b, b^2, b^3 for $n = 1, 2, 3, 4$, respectively, and the population size in the nth generation will be b^{n-1}. In order to deduce a formula for $S(n)$, you can use the three-stage approach that worked well before. First, express $S(n+1)$ in terms of $S(n)$ in two different ways; then set them equal so as to form an equation in $S(n)$; and finally solve this equation.

(i) The sum, $S(n)$, up to the nth generation, is

$$S(n) = 1 + b + b^2 + b^3 + \cdots + b^{n-1}.$$

Now, expressing $S(n+1)$ in terms of $S(n)$ in two ways:

$$S(n+1) = \text{the sum up to the } n\text{th generation} + b^n$$
$$= S(n) + b^n,$$

and

$$S(n+1) = 1 + b + b^2 + b^3 + \cdots + b^{n-1} + b^n$$
$$= 1 + b\left(1 + b + b^2 + b^3 + \cdots + b^{n-1}\right)$$
$$= 1 + bS(n).$$

(ii) Since the right-hand sides of the two preceding formulas are both equal to $S(n+1)$, the right-hand sides themselves must also be equal to each other. So

$$1 + bS(n) = S(n) + b^n.$$

(iii) This is an equation in $S(n)$. To solve it, subtract $S(n)$ and 1 from both sides to get

$$bS(n) - S(n) = b^n - 1.$$

Taking out the factor $S(n)$ gives

$$S(n)(b - 1) = b^n - 1.$$

Finally, divide by $(b - 1)$ to obtain

$$S(n) = \frac{b^n - 1}{b - 1}, \tag{5}$$

which is true for $n = 1, 2, 3, \ldots$ and $b \neq 1$.

This step is not valid for $b = 1$ as you cannot divide by zero.

Notice that this is a more general version of the formula developed in the case of Queen Calcula's gold pieces. There, $b = 2$, so

$$S(n) = \frac{2^n - 1}{2 - 1} = 2^n - 1.$$

Activity 12   How many were coming from St Ives?

Recall the St Ives riddle from Activity 1(c). How many were coming *from* St Ives?

It is not difficult to deal with the even more general case of a population that has a starting size of a, rather than 1. Then the sizes in the first few generations are $a, ab, ab^2, ab^3, \ldots$ for $n = 1, 2, 3, 4, \ldots$, and the size in the nth generation is ab^{n-1}. So the cumulative sum, $S(n)$, is given by

$$S(n) = a + ab + ab^2 + \cdots + ab^{n-1}$$
$$= a(1 + b + b^2 + \cdots + b^{n-1}).$$

Activity 13   A more general formula

The general formula $S(n) = a(1 + b + b^2 + \cdots + b^{n-1})$ is usually written as

$$S(n) = a\left(\frac{b^n - 1}{b - 1}\right). \tag{6}$$

Explain why this is so.

The general formula for $S(n)$ will now be applied to one of the examples you have already looked at—the simplified greenfly population.

Example 4 *Adding up greenfly*

Suppose that instead of dying after producing three offspring on the day after its birth, each greenfly survives to carry on eating the roses, though it produces no more offspring. How many greenfly will be eating your roses on day 5 if the initial population size is 17?

Solution

Initial population size, $a = 17$,

growth factor, $b = 3$,

number of generations, $n = 5$.

So, using formula (6),

$$S(n) = a\left(\frac{b^n - 1}{b - 1}\right)$$

$$= 17\left(\frac{3^5 - 1}{3 - 1}\right)$$

$$= 17\left(\frac{242}{2}\right)$$

$$= 2057.$$

Therefore, on day 5 there could be a total of 2057 greenfly munching your roses if you do not intervene.

Now consider what sort of value the general formula (6) will produce when the growth factor, b, lies between 0 and 1 (that is, when the population is decreasing exponentially). Since b lies between 0 and 1, it follows that b^n will also lie between 0 and 1. Hence the two components of formula (6) that are inside the brackets, $b^n - 1$ and $b - 1$, will both be negative. Since one divides the other, the result will be positive, as is the case when b is greater than 1. However, when b is between 0 and 1, it is usually more convenient to multiply the top and bottom of the fraction by $^-1$ and so write the formula as

$$S(n) = a\left(\frac{1 - b^n}{1 - b}\right). \tag{7}$$

You can see an application of this formula in the next example.

Example 5 *Student power*

A particular mathematics course attracted 4200 students in its first year of presentation. Over the eight-year life of the course, student enrolment fell by 5% each year. How many students in total studied the course?

Solution

One solution would be to calculate the number of students each year—4200, $4200 \times 95\%$, ... and so on—and add them up. But, as this is a case of exponential decrease, the total number of students, $S(n)$, who studied the course can be calculated directly by using formula (7):

$$S(n) = a \left(\frac{1 - b^n}{1 - b} \right).$$

Initial size of student population in year 1, $a = 4200$,

growth factor, $b = 0.95$,

number of years, $n = 8$.

So,

$$S(n) = 4200 \left(\frac{1 - 0.95^8}{1 - 0.95} \right)$$

$$= 28\,273 \text{ (to the nearest whole number)}.$$

Therefore, the total number of students who studied the course over its eight-year life was $28\,273$.

The formula for $S(n)$, in the form of equation (6), can be useful in financial problems. A savings account that earns compound interest has already been discussed. The converse of this situation is a repayment mortgage or a bank loan, where the individual *pays* interest rather than receives it. However, there is an added complication in the calculation here—the borrower makes repayments to the lender (a building society or a bank), as well as paying interest on the loan. The mathematical implications of this are explored in the next example.

Example 6 *A loan with annual repayments*

If you take out a loan, you borrow a sum of money from a building society or a bank and, by making a series of regular repayments, aim to pay off the loan plus interest charges.

Repayments are most often made monthly, but in this example, for convenience, annual repayments are assumed.

Suppose that you borrow £5000 at an annual compound interest rate of 5%. Also suppose that at the end of every year, immediately after the interest has been added, you repay a sum of £300. How much is still owed at the end of every year, after the repayment has been made? Find a formula that can be used to calculate the debt in any year without having to follow the history of the account year by year.

Solution

If you consider what happens to the account in the first few years, you may be able to identify the general rule by which it changes, and then conjecture the required formula. So, adopt the strategy outlined on page 23:

- try out the formula with simple numbers;
- present the results clearly and systematically in a table;
- look for a pattern.

Let the debt (in £) at the end of the nth year be $D(n)$.

Then, at the end of the *first* year (after the repayment of the first £300),

$$D(1) = 5000 \times 1.05 - 300.$$

At the end of the *second* year, the debt is this amount multiplied by 1.05 (for the interest accrued), less 300 (for the repayment). Thus

$$D(2) = (5000 \times 1.05 - 300) \times 1.05 - 300.$$

Because you will be looking for a pattern, do not simplify any further at this stage.

This can be rearranged into a form in which it is easier to see the general pattern:

$$\begin{aligned} D(2) &= 5000 \times 1.05^2 - 300 \times 1.05 - 300 \\ &= 5000 \times 1.05^2 - 300(1.05 + 1) \\ &= 5000 \times 1.05^2 - 300(1 + 1.05). \end{aligned}$$

It is best *not* to simplify any further here.

At the end of the *third* year,

$$\begin{aligned} D(3) &= [5000 \times 1.05^2 - 300(1 + 1.05)] \times 1.05 - 300 \\ &= 5000 \times 1.05^3 - 300(1.05 + 1.05^2) - 300 \\ &= 5000 \times 1.05^3 - 300(1 + 1.05 + 1.05^2). \end{aligned}$$

Summarizing these results in a table gives:

Table 2

Year, n	Debt, $D(n)/£$
1	$5000 \times 1.05 \quad - 300$
2	$5000 \times 1.05^2 - 300(1 + 1.05)$
3	$5000 \times 1.05^3 - 300(1 + 1.05 + 1.05^2)$

▶   Can you see a pattern emerging?

Each successive year, 5000 (the initial loan) is multiplied by a further factor of 1.05, while the terms in the brackets after the 300 have an extra power of 1.05 added to them. At the end of the nth year, the debt will be

$$D(n) = 5000 \times 1.05^n - 300(1 + 1.05 + 1.05^2 + \cdots + 1.05^{n-1}).$$

The second part of this expression is an exponential series with $a = 300$ and $b = 1.05$. Using the formula $S(n) = a\left(\dfrac{b^n - 1}{b - 1}\right)$ for the sum of such series, the debt after n years can be expressed as

$$D(n) = 5000 \times 1.05^n - 300 \times \left(\frac{1.05^n - 1}{1.05 - 1}\right).$$

This can be rearranged and simplified as follows:

$$D(n) = 5000 \times 1.05^n - 300 \times \frac{(1.05^n - 1)}{0.05}$$

$$= 5000 \times 1.05^n - 6000 \times (1.05^n - 1) \qquad \left(\text{since } \frac{300}{0.05} = 6000\right)$$

$$= 5000 \times 1.05^n - 6000 \times 1.05^n + 6000 \qquad \text{(expanding the brackets)}$$

$$= 6000 - 1000 \times 1.05^n \qquad \text{(collecting multiples of } 1.05^n\text{)}.$$

You can see from this formula that the debt decreases year by year. This is what you would expect because, if the loan is eventually to be paid off, the amount repaid must be greater than the amount of interest owed every year.

Using the formula obtained above, it is possible to work out how long it will take until the debt is fully repaid. In other words, when does $6000 - 1000 \times 1.05^n = 0$? Solving this equation gives

$$6000 = 1000 \times 1.05^n$$

$$6 = 1.05^n.$$

So you need to find the value of n for which 1.05^n first exceeds 6. As you can check with your calculator, $1.05^{36} = 5.8$, while $1.05^{37} = 6.1$ (both figures are given to 1 decimal place). This means that the debt would be fully paid off at the end of the 37th year (with a reduced final payment).

Activity 14   A loan again

Suppose you decide to borrow £1000 in an account that charges compound interest at a fixed annual rate of 4% for five years. A feature of this account is that you repay £100 at the end of each year. Derive and simplify a formula for $D(n)$, the outstanding debt (in £) at the end of n years.

What will be the total sum you owe at the end of the five years?

The ideas developed in Example 6 can be applied in a very different context—irrigation.

In countries where rainfall is scarce, the irrigation of fields is an essential part of agriculture. But supplies of water may be limited, and unrestricted irrigation may not be possible. After each period of irrigation, water is lost from the soil by evaporation. The following example deals with a hypothetical simplified irrigation scheme, and examines how the water content of the soil is affected by the gain through nightly irrigation and the loss due to daily evaporation.

Example 7 *Irrigation model*

Consider an irrigation scheme in which farmers are allowed to irrigate their fields from 6 p.m. to 6 a.m. only, applying the *same* maximum amount of water—the daily ration—to their fields every night. Make the simplifying assumption that during the day (that is, from 6 a.m. to 6 p.m.), the strength of the Sun is such that half the total water in the topsoil is lost through evaporation. How does the amount of water in the topsoil vary day by day?

Solution

Let $w(n)$ be the amount of water in the topsoil at the end of the nth day (a day, for this purpose, being a 24-hour period beginning at 6 a.m.). Suppose that irrigation begins after a period of drought when the topsoil is bone dry. Then, on the first day of irrigation, the fields remain dry until 6 p.m., but during the night one daily ration of water is supplied, so by the following morning the water content is 1 (using the constant daily ration as the unit of measurement), assuming that no evaporation occurs between 6 p.m. and 6 a.m. Thus, at the end of day 1,

$$w(1) = 1.$$

During the second day, for the first 12 hours, evaporation reduces the water content by half, so the water content reduces to half a ration; but in the following 12 hours another ration of water is supplied, so the water content goes up to $1 + \frac{1}{2}$. Thus, at the end of day 2,

$$w(2) = 1 + \tfrac{1}{2}.$$

In the next 12 hours, the water content reduces by half to $\frac{1}{2} \times \left(1 + \frac{1}{2}\right) = \frac{1}{2} + \left(\frac{1}{2}\right)^2$, and then a further ration is added, to give

$$w(3) = 1 + \frac{1}{2} + \left(\frac{1}{2}\right)^2.$$

Continuing in the same way, you find that

$$w(n) = 1 + \frac{1}{2} + \left(\frac{1}{2}\right)^2 + \left(\frac{1}{2}\right)^3 + \cdots + \left(\frac{1}{2}\right)^{n-1}.$$

This is the sum of an exponential sequence.

Now, applying formula (7) from page 27, with $b = \frac{1}{2}$ and $a = 1$, you end up with

Note that this version of the formula is used here because b is less than 1.

$$w(n) = \frac{1 - \left(\frac{1}{2}\right)^n}{1 - \frac{1}{2}}$$

$$= 2\left[1 - \left(\frac{1}{2}\right)^n\right]$$

$$= 2 - 2 \times \left(\frac{1}{2}\right)^n.$$

As the days pass, $\left(\frac{1}{2}\right)^n$ becomes extremely small. For instance, for $n = 14$,

$$w(14) = 2 - 2 \times \left(\frac{1}{2}\right)^{14} = 1.999\,878.$$

It follows that after a couple of weeks the water content in the topsoil is very nearly 2. This means that the irrigation scheme reaches a 'steady state', settling on a value of 2 rations of water. Each morning at 6 a.m. there is a double ration of water in the soil. This halves due to evaporation by 6 p.m., when irrigation starts, restoring the water content over the next 12 hours, and the cycle is then repeated. As you will see in Activity 15, if the soil starts off with some moisture in it, then there is a transition to the same steady state but it occurs more quickly.

Here is an alternative way of reaching the conclusion that the steady state is always 2 rations of water. Note that the amount of water in the soil at 6 a.m. on day $n + 1$ is half the amount of that on the previous day, plus 1 ration of water supplied by irrigation. Therefore,

$$w(n + 1) = \tfrac{1}{2}w(n) + 1$$

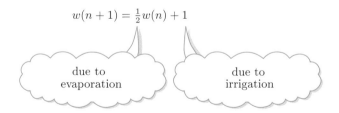

due to evaporation   due to irrigation

However, at the steady state, the amount of water in the soil at 6 a.m. on day n is the *same* as that at 6 a.m. on day $n + 1$. So

$$w(n + 1) = w(n).$$

Call this amount S.

Then the relationship $w(n + 1) = \tfrac{1}{2}w(n) + 1$ can be rewritten as

$$S = \tfrac{1}{2}S + 1.$$

Subtracting $\frac{1}{2}S$ from both sides gives

$\frac{1}{2}S - 1,$

or

$S = 2.$

This confirms the earlier result.

Activity 15   *If the soil isn't dry to begin with* ...

Suppose that the farmer starts off, not with dry soil, but with soil that already contains one third of a day's ration. How does this affect the result of the irrigation programme?

In this section, you have met several situations in which a population changes exponentially. The focus has not just been on the size of the population in any one generation, but on the cumulative sum over several generations. There are two useful formulas for this sum:

$$S(n) = a \left(\frac{b^n - 1}{b - 1} \right) \quad \text{if } b > 1,$$

$$S(n) = a \left(\frac{1 - b^n}{1 - b} \right) \quad \text{if } 0 < b < 1.$$

Outcomes

After studying this section, you should be able to:

◇   decide, in a given situation involving exponential change, when it is appropriate to calculate the sum over several generations (Activities 11, 12, 14 and 15);

◇   understand the derivation and use of the formulas for the cumulative sum (Activity 11–15).

3   Exponential graphs

 Aims   This section aims to help you become familiar with the graphs of exponential functions. It also introduces the number denoted by e.   ◇

In this section you will use your calculator to plot different forms of exponential functions. In Section 1 you dealt with two such forms, $P(n) = b^n$ and $P(n) = ab^n$. Shortly, you will look at the shape of these curves and will also consider what happens when a constant term, c, is added, giving $P(n) = ab^n + c$. The latter form is the most general way of expressing an exponential function.

You will also study a very special exponential function, known as *the* exponential, based on a particular number represented by the letter e.

3.1   Investigating exponential graphs

A feature of your calculator is that functions can only be entered in a standard notation that uses the letters x and y, rather than, say, n and $P(n)$. Thus the equation $P(n) = b^n$ would need to be rewritten as $y = b^x$. Correspondingly, the equation $P(n) = ab^n$ would be entered as $y = ab^x$.

Another form of exponential function was introduced in Example 7. There the formula used to model the amount of water, $w(n)$, in the topsoil at the end of the nth day of irrigation was given as

$$w(n) = 2 - 2 \times \left(\tfrac{1}{2}\right)^n .$$

Replacing n by x and $w(n)$ by y gives

$$y = 2 - 2 \times \left(\tfrac{1}{2}\right)^x .$$

Rather than being of the simple exponential form $y = b^x$ or $y = ab^x$, this function takes the more general exponential form $y = ab^x + c$, where a, b and c are constants. In the irrigation example,

$$a = {}^-2, \qquad b = \tfrac{1}{2} \qquad \text{and} \qquad c = 2.$$

Activity 16   *Conversion of formulas*

Here are some of the formulas derived in the previous two sections. Convert each of them into the general form $y = ab^x + c$ by suitable substitutions of x and y. Then identify the particular values of a, b and c.

(a) $P(n) = 3 \times \left(\tfrac{4}{3}\right)^n$ (b) $P(n) = 5^n$

(c) $S(n) = 2^n - 1$ (d) $D(n) = 6000 - 1000 \times 1.05^n$

The rest of this section consists of activities in which you will use your calculator to investigate the shapes and properties of the graphs of various exponential functions of the form $y = ab^x$ or $y = ab^x + c$. Section 12.1 of the *Calculator Book* provides all the information that you will need about entering, tabulating and graphing exponential functions.

Now study Section 12.1 of Chapter 12 in the Calculator Book

In the situations described in Sections 1 and 2 of this unit, the discrete variable, n, represented a number of stages, steps or generations. As such, n could sensibly only be one of the integers 0, 1, 2, 3.... By contrast, an important feature of all the exponential graphs that you plotted in Section 12.1 of the *Calculator Book* was that they were *continuous* functions. In other words, rather than being restricted to discrete stages or steps or generations, each graph was able to take a value even when x was not a whole positive number. Representing growth as a continuous function in this way will be looked at more closely in Section 4.

To get a better sense of the features of exponential graphs, it is useful to compare them with the graphs produced by other kinds of functions. In general, graphs belong to families (linear, quadratic, exponential, ...), and each family has its own characteristic shape. So what are the characteristics of an exponential graph?

Activity 17 *An exponential graph*

In Section 12.1 of the *Calculator Book*, you plotted the exponential function $y = 3^x$ for x in the range $^-3$ to 3 and y in the range 0 to 28.

(a) Describe the position and general shape of this exponential graph, making reference to its gradient.

(b) Compare this graph with a straight-line graph and with a graph based on a quadratic function—the parabola. In which ways (if any) is the shape of the exponential curve similar to either of these? In which ways is it different?

The observations you made in part (a) of Activity 17 can be confirmed by considering some numerical values of the function $y = 3^x$.

Activity 18 *Numerical values*

Obtain a table of values of $y = 3^x$ on your calculator and then use the table to answer the following questions:

(a) What is the value of y when $x = 0$? When $x = 1$? When $x = ^-1$?

(b) Describe how the value of 3^x changes as x increases, starting at $x = 0$, and as x decreases (that is, becomes more negative), starting at $x = 0$.

Now take the idea of comparing linear, quadratic and exponential graphs further.

Activity 19   *Making comparisons*

From part (a) of Activity 18, you know that the exponential function $y = 3^x$ satisfies $y = 1$ when $x = 0$, and $y = 3$ when $x = 1$. A straight line that also satisfies these conditions has the equation

$$y = 2x + 1.$$

The equation of a parabola that satisfies the conditions is

$$y = 2x^2 + 1.$$

Plot the functions $y = 2x + 1$ and $y = 2x^2 + 1$ on the same screen as the exponential $y = 3^x$.

(a) Write down one significant similarity between the exponential graph and the straight line, and one significant difference (other than that the exponential is not straight).

(b) Write down one significant similarity between the exponential graph and the parabola, and one significant difference.

There is a convenient bit of terminology to describe the fact that 3^x gets closer and closer to 0 as x get smaller and smaller (that is, as x becomes more and more negative): you can say that 3^x *tends to 0 as x tends to minus infinity* ($^-\infty$). Any straight line that a curve approaches in this way but never meets is known as an *asymptote*. Thus the exponential function $y = 3^x$ has the x-axis as an asymptote.

The word *asymptote* is derived from three Greek words: *a-* not, *sum-* together and *ptotos-* falling.

The plus sign is optional here, just as it is optional in front of a positive number; it is included to draw attention to the contrast between this and the previous case.

The behaviour of 3^x as x gets bigger and bigger can be expressed in similar terms: you can say that 3^x *tends to plus infinity* ($+\infty$) *as x tends to plus infinity* ($+\infty$). However, there is no asymptote here.

Activity 20   *Tendencies*

(a) Describe the behaviour of the linear function $y = 2x + 1$ as x tends to $+\infty$, and as x tends to $^-\infty$. (Your answer should take the form 'As x tends to $+\infty$, $2x + 1$ tends to $\ldots$'.)

(b) Do the same for the quadratic function $y = 2x^2 + 1$.

The tasks that you carried out in Activity 20 have a rather wider purpose than you may have realized. It is often useful to have an idea of the overall shape of a graph. A good strategy in this regard is to picture what sort of values the graph is *tending to* at the left and right extremes (that is, for large negative and large positive values of x). If you can remember these different 'tendencies', you should readily be able to distinguish the graphs of the three types of function—linear, quadratic and exponential.

The next step is to compare the graphs of exponentials that have different bases, b.

Activity 21   *The characteristic shape of exponential graphs*

Use your calculator to plot the graphs of $y = 3^x$, $y = 4^x$ and $y = 5^x$ on the same screen, with x in the range $^-1.5$ to 1.5 and y in the range $^-1$ to 4.

(a) For each of these graphs, what is the value of y when $x = 0$? When $x = 1$?

(b) Where do you think the graph of $y = 6^x$ would lie in relation to the others? What about $y = 2^x$? And what about $y = 3.5^x$? First, decide what answers you expect, and then confirm them by using the calculator to plot the graphs.

(c) Can you make a general statement about the shape of the graph of $y = b^x$ when b is some number greater than 1

   If $B > b$ and both B and b are greater than 1, how would the graphs of $y = b^x$ and $y = B^x$ compare?

Activity 22   *Investigating values of b between 0 and 1*

Plot the following functions on your calculator one at a time, and then all together: $y = 0.5^x$, $y = 0.25^x$, $y = 0.125^x$. Use the same x- and y-ranges as in Activity 21.

(a) In each case, what is the value of y when $x = 0$? When $x = 1$? When $x = {}^-1$?

(b) Describe the common shape of all three graphs. How do the three graphs differ?

   Where do you think the graph of $y = 0.6^x$ would lie in relation to the others? What about $y = 0.2^x$? And what about $y = 0.025^x$? Decide what answers you expect, and then confirm them by using the calculator.

(c) What would you expect to be the general shape of the graph of $y = b^x$ when b is some positive number less than 1? In this case, what do you think b^x tends to as x tends to $+\infty$, and as x tends to $^-\infty$?

If $B > b$ and both B and b are between 0 and 1, how would the graphs of $y = b^x$ and $y = B^x$ compare?

(d) Is there any relation between the graphs of equations of the form $y = B^x$ when $B > 1$, and the graphs of equations of the form $y = b^x$ when b is between 0 and 1? If so, what is the relation?

You have seen in Activities 21 and 22 that the graph of the simple exponential function $y = b^x$ curves upwards to the right for values of the base b greater than 1, and shows the classic 'decay' curve downwards to the right for values of b between 0 and 1. But what occurs when $b = 1$?

Activity 23   A common point of exponential graphs

(a) Describe the graph of $y = 1^x$. You may use your calculator if you really want to.

(b) There is only one point that *every* exponential graph $y = b^x$ passes through. What is it?

So far you have considered graphs of functions like $y = b^x$; these represent the special case of the exponential function $y = ab^x$ when $a = 1$. It is now time to find out what effect the value of the parameter a has on the graph.

Activity 24   The effect of changing the value of a

Plot the function $y = 2^x$ on your calculator, using the same x- and y-ranges as in Activities 21 and 22. Then superimpose, one at a time, the graphs of $y = 3 \times 2^x$, $y = 4 \times 2^x$ and $y = 0.3 \times 2^x$.

(a) In each case, what is the value of y when $x = 0$?

(b) The graphs $y = 3 \times 2^x$ and $y = 4 \times 2^x$ are members of the family $y = a \times 2^x$. Describe the similarities and the differences between them.

Activity 25   Multiplying by $^-1$

(a) Plot the function $y = {}^-2^x$; that is, $y = (^-1) \times 2^x$.

(b) This graph is one of the family $y = a \times 2^x$. How does it relate to other members of this family?

You have been looking at graphs of functions of the form $y = ab^x$; these are members of the wider family of functions $y = ab^x + c$ when $c = 0$. So, next turn your attention to graphs of functions of the form $y = ab^x + c$.

Activity 26   *Adding or subtracting a constant term*

(a) Can you predict how the graph of $y = 2^x - 1$ is related to the graph of $y = 2^x$? (The relationship is the same as that between $y = x^2 - 1$ and $y = x^2$, which you investigated in *Unit 11*.) Check by plotting both functions on your calculator.

(b) Predict the graph of $y = 2^x + 1$ and check it on the calculator.

(c) Plot the function $y = 6000 - 1000 \times 1.05^x$ for x between 0 and 40.

This corresponds to the formula for the outstanding debt in Example 6, but with y representing the debt and x the relevant year. Find when the debt reduces to £0.

You have already seen a function of the form $y = ab^x + c$ for which b lies between 0 and 1. It arose in the model of irrigation in Example 7, and was $y = 2 - 2 \times \left(\frac{1}{2}\right)^x$. The value of the water content of the soil (denoted now by y) very quickly gets close to 2 and stays there (referred to as reaching a 'steady state'). This behaviour is clearly shown when the graph of the function is plotted, as you will see in the next activity.

Activity 27   *Plotting the irrigation model*

Plot $y = 2 - 2 \times \left(\frac{1}{2}\right)^x$ for x between 0 and 10. Comment on the behaviour of the function as x tends to $+\infty$.

Activity 27 has illustrated an important property of functions of the form $y = ab^x + c$ when b lies between 0 and 1; in such cases, $ab^x + c$ tends to c as x tends to $+\infty$. The reason is that b^x tends to 0 as x tends to $+\infty$. This kind of function, therefore, can be used to model something that grows (or decays) towards a limit, c.

You have now considered the behaviour of the general exponential function $y = ab^x + c$ when $b > 1$ (the condition of unlimited growth) and when $0 < b < 1$ (the condition of decline to a limiting value of c). But what happens when $b = 1$? This produces the function $y = a + c$, which is a horizontal line positioned $a + c$ units above the x-axis. However, although this is technically a special case of the exponential function, it would not normally be referred to as such.

You might like to update your Handbook notes to include information about the functions $y = ab^x$ and $y = ab^x + c$. For a summary of the behaviour of these functions, see the table on page 42.

3.2   The exponential

As you have seen, the value of the base b plays a crucial role in defining an exponential function. There is one value of b that is particularly important. It is referred to as e. When $b = e$, the simple exponential function $y = b^x$ becomes $y = e^x$; this has special properties and is often referred to as *the* exponential function.

In Section 3.1, it was shown that the main difference between graphs of $y = b^x$ for different values of b is in their steepness: the larger b is, the steeper the graph is. However, all such graphs go through the point $(0, 1)$. When $b = 1$, the graph is a horizontal straight line and so its gradient is 0. As b increases from 1, the gradient of the graph of $y = b^x$ at the point $(0, 1)$ gets larger and larger. When $b = 10$, say, the gradient at that point is much bigger than 1. There is some value of b between 1 and 10, therefore, for which the gradient of $y = b^x$ at the point $(0, 1)$ is exactly 1.

It turns out that the value of b for which the corresponding exponential function has a gradient of 1 at $(0, 1)$ is the number e; this number is approximately $2.718 \ldots$. In fact, not only is the gradient of *the* exponential function $y = e^x$ equal to the value of y at the point $(0, 1)$, but this is true for *every* point on the curve: for all values of x, the value of y is equal to the corresponding gradient.

This special property of $y = e^x$ can be explored by using the calculator. Notice that e^x is important enough to warrant a special place (as a second function) on the calculator keyboard. In the next activity you are asked to focus on just one point, $(0, 1)$, and to find out whether the y value, which is 1, is equal to the slope of the curve at that point. Later, using the *Calculator Book* as a guide, you will check whether $y = e^x$ has gradient values that are equal to y for *all* values of x.

Activity 28   *The gradient of the exponential*

To do this, use the calculator command **ZSquare**.

Plot the function $y = e^x$ in a small region around the point $(0, 1)$, making that point the centre of your screen, and setting up the window so that the units on the x- and y-axes are of equal length on the screen (to ensure that lines of gradient 1 appear at 45° to the axes). Next, zoom in two or three times, keeping $(0, 1)$ at the centre of the screen. At each zoom, the graph appears more like a straight line. This is a general feature of zooming in on graphs. The point of interest in this case is that the line has gradient 1. You can check by using the trace facility.

When you have zoomed in sufficiently often that the graph looks straight, trace along the graph one pixel to the right of $(0, 1)$, and note the x and y coordinates. You should find that both coordinates have changed by virtually the same small amount. Repeat this by returning to $(0, 1)$, zooming in further and again tracing one pixel to the right.

In Section 12.1 of the *Calculator Book*, you met the inverse of the function $y = 10^x$; this was the logarithm function to base 10. Just as you can use any base b to form an exponential such as $y = b^x$, so you can form logarithms to any base b, as in $y = \log_b x$. The two functions $y = b^x$ and $y = \log_b x$ are inverses of one another: each undoes what the other one does. So $y = \log_{10} x$ and $y = 10^x$ are inverses of one another. Similarly, $y = \log_2 x$ and $y = 2^x$ are inverses of one another.

Corresponding to *the* exponential, $y = e^x$, there is a particular logarithm function $y = \log_e x$, which is referred to as 'logarithm to the base e'. The expression $\log_e x$ is often denoted by $\ln x$, which stands for 'natural logarithm'. It follows that $y = \ln x$ is the inverse of $y = e^x$. You will find that e^x and ln share a key on the calculator for this reason.

The exponential and the natural logarithm function are often the preferred choices when using exponentials for theoretical purposes in mathematics and science, and you are likely to come across them again if you take your mathematics further.

There is a more detailed treatment of logarithms in Section 4.3.

At this point, clarify your understanding of *the* exponential function and its inverse, the natural logarithm, by exploring both on your calculator.

Now study Section 12.2 of Chapter 12 in the Calculator Book

The conclusion of the last exercise in Section 12.2 of the *Calculator Book* was that the gradient of $y = e^x$ *at any point* is equal to the value of y at that point. In other words, the graph of the function $y = e^x$ and of its gradient function are identical. This is the defining property of *the* exponential function, and is important to remember.

In some respects, *the* exponential is rather like π: in particular, the decimal expansion of e, like that of π and $\sqrt{2}$, neither terminates nor recurs. This means that e, like π, cannot be represented exactly as a fraction, since the decimal expansion of any fraction either terminates (like $\frac{1}{2}$ and $\frac{1}{4}$) or is repeating (like $\frac{1}{3}$ and $\frac{1}{7}$). Numbers like this are called *irrational*. A rational number is one that can be expressed as a fraction, or a *ratio* of two whole numbers. An irrational number, such as $\sqrt{2}$, e and π, is one that cannot. (In case you are moved to protest that $\pi = \frac{22}{7}$, remember $\frac{22}{7}$ is only *an approximation* to the value of π, though it is quite a good one.)

'Ir-rational' means 'without ratio'.

You will meet e again in Section 4.4.

41

In the present section the focus has been on the properties of exponential graphs. For convenient reference, these properties are summarized in the table below.

Table 3

Function	Condition	Characteristics of graph
$y = b^x$	$b > 1$	The graph curves upwards to the right, passing through $(0, 1)$.
	$0 < b < 1$	The graph curves downwards to the right (that is, it decays), passing through $(0, 1)$.
	$b < 0$	The function is not defined for negative values of b.
$y = ab^x$	$b > 1$	For $a > 0$, the graph curves upwards to the right, passing through $(0, a)$. For large values of a, the curve is steep. For small values of a, the curve is flatter.
		For $a < 0$, the curve becomes the mirror image (reflected in the x-axis) of the graph where a is positive.
	$0 < b < 1$	For $a > 0$, the graph curves downwards to the right (that is, it decays), passing through $(0, a)$.
		For $a < 0$, the curve becomes the mirror image (reflected in the x-axis) of the graph where a is positive.
	$b < 0$	The function is not defined for negative values of b.
$y = ab^x + c$		The graph is identical to that of $y = ab^x$ except that it is shifted upwards by an amount c.
$y = e^x$		This is a special case of $y = b^x$ where b is given a particular numerical value: $2.71828\ldots$. This number is referred to as e.
		The function $y = e^x$ has the very special property that, at any point on the graph, the gradient of the curve is equal to the value of the function at that point.

Outcomes

After studying this section, you should be able to:

◇   convert exponential functions into the general form $y = ab^x + c$ (Activity 16);

◇   describe the characteristic features of the graphs of exponential functions (Activities 17–20);

◇   describe how the values of the constants a and b affect the graph of $y = ab^x$ (Activities 21–25);

◇   describe the graph of $y = ab^x + c$ and draw conclusions about the long-term behaviour of a population when the growth factor is between 0 and 1 (Activities 26 and 27);

◇   describe the graph of *the* exponential function $y = e^x$ and its gradient function (Activity 28 and *Calculator Book* Exercises 12.4–6).

4 Exponents

Aims   This section introduces some important rules for operating with and simplifying expressions that involve exponents.   ◇

The graphical investigations in Section 3 brought to light some interesting features of the graphs of functions of the form $y = ab^x$, notably that ab^x has a continuous graph and so b^x must have a value even when x is not a whole positive number. In this section, you will see some properties of exponents (or powers) that should explain this 'continuous' feature of the graph.

4.1 Rules for exponents

Some of the basic ideas in this section were covered in *Preparatory Resource Book A*, Module 4.

A kilometre is 1000 metres; a kilogram is 1000 grams. The prefix 'kilo-' means 'a thousand', and putting 'kilo' in front of the name of a unit produces the name of a new unit that is 1000 times the previous one. This is a principle that could be applied over and over again: you might call 1000 kilometres a kilokilometre, and 1000 kilokilometres a kilokilokilometre. But, in fact, there are special names for these large units: a kilokilo(some unit) is called a mega(some unit), and a kilokilokilo(some unit) is a giga(some unit).

▶   How many metres are there in a gigametre?

A kilometre is 1000 metres, or 10^3 metres, so a megametre is $1000 \times 1000 = 1\,000\,000$ metres, or 10^6 metres, and a gigametre is $1000 \times 1\,000\,000 = 1\,000\,000\,000$ metres, or 10^9 metres. Notice how much more convenient it is to write these large numbers in exponential (or scientific) form than in ordinary decimal form. This is partly because the exponential form saves space: but there is more to it than that. There is a simple pattern in the way that the exponent changes each time the number is multiplied by 1000, or 10^3: the exponent just increases by 3.

So,

$$10^3 \times 10^3 = 10^{3+3} = 10^6,$$
$$10^3 \times 10^6 = 10^{3+6} = 10^9.$$

This suggests a general rule that applies whenever two numbers, expressed as powers of 10, are multiplied together. The rule is

$$10^m \times 10^n = 10^{m+n}.$$

To take a particular example, think about $10^3 \times 10^6 = 10^9$:

$$10^3 \times 10^6 = (10 \times 10 \times 10) \times (10 \times 10 \times 10 \times 10 \times 10 \times 10)$$
$$= 10 \times 10 \times 10 \times 10 \times 10 \times 10 \times 10 \times 10 \times 10$$
$$= 10^9.$$

All you have to do is count the number of 10s:

$$10^m \times 10^n = \underbrace{(10 \times 10 \times \cdots \times 10)}_{m \text{ tens}} \times \underbrace{(10 \times 10 \times \cdots \times 10)}_{n \text{ tens}}$$
$$= \underbrace{10 \times 10 \times \cdots \times 10}_{(m+n) \text{ tens}}$$
$$= 10^{m+n}.$$

Activity 29   *Some other ways of looking at multiplication*

(a) Think about multiplying together two numbers of the form $100\ldots000$: for instance, $10\,000\,000 \times 100\,000\,000\,000$. How can you tell how many zeros there will be after the 1 in the answer?

(b) A million is a thousand thousands; a billion is a thousand millions. Express as powers of 10: ten billion millions; one hundred billion billions; one thousand million billion million billions.

The rule for calculating the product of powers is commonly used when multiplying numbers expressed in scientific notation. For example, the mean distance from the Earth to the Sun is (approximately) 1.5×10^8 kilometres. To express this in metres you merely have to multiply it by 10^3: this means you add 3 to the exponent 8, to obtain 1.5×10^{11} metres.

Take another example: the speed of light is about 3×10^5 kilometres per second, and there are approximately 3×10^7 seconds in a year, so light travels $(3 \times 10^5) \times (3 \times 10^7)$ kilometres in a year. To evaluate this product, first multiply the two factors 3 together to get $9 \times 10^5 \times 10^7$; then add the exponents 5 and 7. Hence the distance light travels in one year is about 9×10^{12} kilometres. This distance is usually called a *light-year*.

Activity 30   *Inching to the Sun*

(a) The mean distance from the Earth to the Sun is 1.5×10^8 kilometres. There are about 3.9×10^4 inches in a kilometre. What is the mean distance from the Earth to the Sun in inches?

(b) The mean distance of the planet Pluto from the Sun is about 5.9×10^9 kilometres. (Pluto is the planet that is furthest from the Sun.) How far is it from Pluto to the Sun in inches?

You have looked at the rule for multiplying exponentials. But what is the rule that operates when one exponential is divided by another, or an exponential is itself raised to a power? These questions are explored in the *Calculator Book*.

Now study Section 12.3 of Chapter 12 in the Calculator Book.

Table 4 summarizes the rules for calculating with exponentials. Note that the rules have been expressed for *any* base number b, rather than just for a base of 10.

Table 4

Calculation	In symbols	In words	Example
Multiplying exponentials that have the same base.	$b^m \times b^n = b^{m+n}$	When multiplying, add the powers.	$6^4 \times 6^3 = 6^{4+3}$ $= 6^7$
Dividing exponentials that have the same base.	$b^m \div b^n = b^{m-n}$	When dividing, subtract the powers.	$6^7 \div 6^3 = 6^{7-3}$ $= 6^4$
Raising an exponential to a power.	$(b^m)^n = b^{m \times n}$ $= b^{mn}$	When raising to a power, multiply the powers.	$(6^2)^3 = 6^{2 \times 3}$ $= 6^6$

You might like to add these rules to your Handbook notes.

So far you have considered positive whole-number exponents. Turn your thoughts now to the possible interpretation of an exponential in which the exponent is zero or a negative whole number. What meaning can be attached to numbers like 5^0 or 6^{-3}? These questions will be investigated here using simple numerical examples.

▶ What is the meaning of a *zero exponent*?

Beware of confusing 5 raised to the power of zero (5^0) with the angle 5 degrees ($5°$).

Consider the calculation $5^4 \div 5^4$. Applying the rule for dividing exponentials gives $5^4 \div 5^4 = 5^{4-4} = 5^0$. However, any number divided by itself is equal to 1. So, $5^0 = 1$. This result would be true not just for 5 but for any base number b. In general, then,

$$b^0 = 1.$$

▶ What is the meaning of a *negative exponent*?

Consider the calculation $6^2 \div 6^5$. The rule for dividing exponentials gives $6^2 \div 6^5 = 6^{2-5} = 6^{-3}$. But if this calculation is set out as

$$\frac{6 \times 6}{6 \times 6 \times 6 \times 6 \times 6},$$

cancelling two of the sixes gives

$$\frac{1}{6 \times 6 \times 6}, \quad \text{or} \quad \frac{1}{6^3}.$$

So,

$$6^{-3} = \frac{1}{6^3}.$$

Again, this result is true not just for 6 but for any base number b. It is also true not just for $^-3$ and 3, but for any exponent ^-x or x. Therefore, in general,

$$b^{-x} = \frac{1}{b^x}.$$

You can also derive this general result in another way, starting from

$$1 = b^0.$$

Divide both sides by b^x:

$$1 \div b^x = b^0 \div b^x$$
$$= b^{0-x}$$
$$= b^{-x}.$$

4.2 Fractional exponents

Section 4.1 dealt with exponents that were integers (whole numbers), including positive and negative values as well as zero. But what meaning can be attached to an exponential like 10^p, when p is a fraction?

As before, this question can be investigated using actual numbers. For instance, what meaning can be attached to numbers like $10^{1/2}$, $10^{1/3}$ and $10^{3/4}$?

▶ What is the meaning of $10^{1/2}$?

Consider the calculation $10^{1/2} \times 10^{1/2}$. According to the rule for multiplying exponentials given in Table 4,

$$10^{1/2} \times 10^{1/2} = 10^{1/2+1/2}$$
$$= 10^1$$
$$= 10.$$

This calculation has demonstrated that $10^{1/2}$ multiplied by itself gives 10. The only positive number that has this property is $\sqrt{10}$. So $10^{1/2} = \sqrt{10}$.

▶   What is the meaning of $10^{1/3}$?

Consider the calculation $10^{1/3} \times 10^{1/3} \times 10^{1/3}$. Again, applying the rule for multiplying exponentials,

$$10^{1/3} \times 10^{1/3} \times 10^{1/3} = 10^{1/3+1/3+1/3},$$
$$= 10^1,$$
$$= 10.$$

The only number with this property is the cube root of 10, that is $\sqrt[3]{10}$. It follows that $10^{1/3} = \sqrt[3]{10}$.

You have seen that $10^{1/2} = \sqrt{10}$ and $10^{1/3} = \sqrt[3]{10}$. The evolving pattern suggests that $10^{1/n} = \sqrt[n]{10}$. To prove this general result, use the rule for raising an exponential to a power (see Table 4). This gives

$$\left(10^{1/n}\right)^n = 10^{(1/n \times n)} = 10^1 = 10.$$

Hence

$$10^{1/n} = \sqrt[n]{10}.$$

In other words, $10^{1/n}$ is the nth root of 10.

That takes care of fractional exponents of the form 'one over something'.

▶   What about more general fractional exponents, such as $\frac{3}{4}$?

Well, $\frac{3}{4} = \frac{1}{4} \times 3$, and once again the rule for raising an exponential to a power determines how to interpret $10^{3/4}$:

$$10^{3/4} = 10^{(1/4 \times 3)} = (10^{1/4})^3 = (\sqrt[4]{10})^3.$$

So, to calculate $10^{3/4}$, take the fourth root of 10 and cube the result.

This rule is valid for any fractional exponent. Therefore,

$$10^{m/n} = 10^{(1/n \times m)} = (10^{1/n})^m = (\sqrt[n]{10})^m.$$

The rules for working with exponentials hold for powers not just of 10 but of any positive base number b—it does not even have to be a whole number. Thus

$$b^{m/n} = (\sqrt[n]{b})^m.$$

It may have occurred to you that there was actually a choice as to how to calculate $10^{3/4}$. The fraction $\frac{3}{4}$ was expressed as $\frac{1}{4} \times 3$, and so $10^{3/4} = (10^{1/4})^3 = (\sqrt[4]{10})^3$. But you could just as well have said $\frac{3}{4} = 3 \times \frac{1}{4}$, which would have given

$$10^{3/4} = 10^{(3 \times 1/4)} = (10^3)^{1/4} = \sqrt[4]{10^3},$$

instead of $(\sqrt[4]{10})^3$. This would imply that you can calculate $10^{3/4}$ by cubing 10 first and then taking the fourth root of the result, rather than taking the fourth root of 10 first and cubing the result. It does not matter in which order you do these two operations, you get the same answer either way. This, too, is true whatever the base number b and whatever the fraction m/n. In general,

$$b^{m/n} = (\sqrt[n]{b})^m = \sqrt[n]{(b^m)}.$$

Activity 31 *Check*

(a) Use your calculator to work out 2^3, and then take the fourth root of the answer. Now work out the fourth root of 2, and cube the answer. Why are the two answers the same?

(b) Use your calculator to find $10^{0.75}$. Now calculate the fourth power of your answer and explain the result.

(c) Calculate the following without using your calculator:

$$8^{2/3}, \qquad 9^{3/2}, \qquad \sqrt[3]{2^6}, \qquad \sqrt[4]{100^2}.$$

One result of your work in this section is that you should now know what your calculator is actually calculating when you enter, say, $17.43^{-2.25}$. You should also understand better why, when you get the calculator to plot $y = 2^x$, you obtain a continuous curve, extending to the negative x-axis, and not just a set of discrete points.

$$17.43^{-2.25} = \frac{1}{17.43^{2.25}}$$
$$= \frac{1}{17.43^{9/4}}$$
$$= \frac{1}{\sqrt[4]{17.43^9}}$$

Table 5 summarizes the rules for zero, negative and fractional exponents. As was the case in Table 4, the rules have been expressed for *any* base number b.

Table 5

Exponent	In symbols	Example
Zero exponent	$b^0 = 1$	$7^0 = 1$
Negative exponent	$b^{-n} = \dfrac{1}{b^n}$	$5^{-2} = \dfrac{1}{5^2} = \dfrac{1}{25}$
Fractional exponent	$b^{m/n} = \left(\sqrt[n]{b}\right)^m = \sqrt[n]{b^m}$	$27^{2/3} = \left(\sqrt[3]{27}\right)^2 = 3^2 = 9$

You might like to add these rules to your Handbook notes.

4.3 Logarithms

You saw in Section 3.2 that the inverse of *the* exponential function $y = e^x$ is the natural logarithm function $y = \log_e x$ or $\ln x$. Logarithms can, in fact, be calculated to any base, but the most common bases are e and 10.

Consider the logarithm of 5, calculated to base 10, which is written as $\log_{10} 5$. It is the power to which 10 must be raised to give the answer 5.

▶   How could you set about finding this power?

One possible way of answering this question is to think of $10^x = 5$ as an equation that can be solved to find the required number, x. To solve such an equation, you could plot the function $y = 10^x$ and find where the graph crosses the line $y = 5$.

Activity 32   *Solving the equation*

Using the method developed in Section 8.5 of the *Calculator Book*, solve the equation $10^x = 5$ graphically on your calculator, giving your answer correct to 3 decimal places.

As you have just found, the value of x for which $10^x = 5$ is 0.699, correct to 3 decimal places. Hence the logarithm of 5 to base 10 is 0.699, or

$$\log_{10} 5 = 0.699 \text{ (to 3 d.p.)}.$$

Activity 33   *Solution check*

Check that $10^{0.699}$ is a reasonably good approximation to 5, by calculating it directly (using your calculator). It is not *exactly* 5, because 0.699 is not the exact solution of the equation $10^x = 5$. Solving the equation graphically by zooming in (as you did in Activity 32) can give you a succession of approximate solutions which get closer and closer to (or which 'tend to' in the language of Section 3) the exact solution, in the sense that the corresponding values of 10^x get closer and closer to (or tend to) 5. To see this process in action, try calculating the values of 10^x when $x = 0.7$, 0.699 and 0.698 97 (these are successive approximations to the solution).

Having solved the equation $10^x = 5$, think about a related problem: how to solve $10^x = 500$.

This equation can be rewritten as

$$10^x = 5 \times 100$$
$$= 5 \times 10^2$$
$$= 10^{0.699} \times 10^2.$$

The rule for multiplying exponentials (that is, $b^m \times b^n = b^{m+n}$) gives

$$10^x = 10^{0.699+2}$$
$$= 10^{2.699}.$$

So

$$x = 2.699 \text{ (to 3 d.p.).}$$

Hence the logarithm of 500 is 2.699, or

$$\log_{10} 500 = 2.699 \text{ (to 3 d.p.).}$$

Solutions to equations of the form '$10^x = $ something' have long been recognized as useful aids in calculation; indeed, until the invention of the calculator, people had to depend almost entirely on tables of logarithms to carry out complicated numerical calculations. What you were doing earlier when you calculated the logarithm of 5 to base 10 was the equivalent of looking up the number 5 in log tables.

Some students may remember using 'log tables' at school in order to carry out more complicated calculations.

Historical note

The word *logarithm*, coined in the seventeenth century by the Scottish mathematician John Napier, comes from two Greek words: *logos*, meaning 'ratio', and *arithmos*, meaning 'number'. So logarithms are 'ratio numbers'.

There are some Reader articles about the history and development of logarithms: now would be a good time to read them, if you are interested.

Activity 34   Check

Using your calculator's log key, confirm that $\log_{10} 5 = 0.699$ (correct to 3 decimal places). If by any chance you can lay your hands on a book of log tables, you might also like to check that the entry for the logarithm of 5 to base 10 is 0.699.

In general, a logarithm to the base 10 is defined as

$$a = \log_{10} b \quad \text{if} \quad 10^a = b. \tag{8}$$

That is, the log of b is the power to which 10 must be raised to give b. Put the other way around,

$$10^a = b \quad \text{if} \quad \log_{10} b = a.$$

Finding a logarithm is simply a matter of 'undoing' the process of forming an exponent: the logarithm and the exponential are related to each other in exactly the same way as, for example, multiplying by 2 and dividing by 2, or raising to the power 3 and taking the cube root. That is to say, if you apply one of these operations to some number, and then apply the other operation to the result, you get back to the number you started with. So, if you first calculate the logarithm of a number to base 10, and then raise 10 to the power of the result, you end up with the original number. Likewise, if you raise 10 to some power, and then take the logarithm to base 10 of the result, you end up with the number that was the power you began with.

Earlier in this section, the expression $\log_{10} 5$ was described as the power to which 10 must be raised to get the answer 5. Similarly, $\log_{10} x$ is the power to which 10 must be raised to get the answer x. Based on this understanding of what a logarithm is, the 'doing–undoing' relationship between logarithms and exponentials can be summarized in a couple of important identities:

$$10^{\log_{10} x} = x. \tag{9}$$

$$\log_{10}(10^x) = x, \tag{10}$$

Activity 35   *Doing and undoing*

(a) Find the value of $10^{2.34}$, and then find the logarithm to base 10 of the result.

(b) Find $\log_{10} 3.5$, and then raise 10 to the power of your answer.

If you have previously met log tables as a calculating aid, you may have been surprised to learn that there is a connection between logarithms and exponentials. In books of log tables there were two types of tables: those for logarithms and those for *antilogarithms*—but no obvious sign of exponentials. However, since the function of antilogarithms is to undo logarithms, *antilogarithms* are *exponentials* under another name.

Activity 36   *Log graphs*

Use your calculator to plot the graphs of $y = \log_{10} x$ and $y = 10^x$ together.

Comment on how these graphs are related.

Using identity (9), you can relate the rules for exponentials to the rules that were used historically for calculating with logarithms. Suppose you want to multiply two numbers together, say x and y. The traditional method for doing this with logarithms was to take the logarithm (to base 10) of each of the numbers, add the two, and take the antilogarithm of the result. A proof that this method is correct is set out below.

From identity (9),

$$x = 10^{\log_{10} x} \quad \text{and} \quad y = 10^{\log_{10} y}.$$

So,

$$x \times y = 10^{\log_{10} x} \times 10^{\log_{10} y}$$
$$= 10^{\log_{10} x + \log_{10} y}.$$

But, also from identity (9),

$$x \times y = 10^{\log_{10}(x \times y)}.$$

Then equating the two expressions for $x \times y$ gives

$$10^{\log_{10}(x \times y)} = 10^{\log_{10} x + \log_{10} y}.$$

Hence

$$\log_{10}(x \times y) = \log_{10} x + \log_{10} y. \tag{11}$$

This means that if you add the logs of x and y, you get the log of their product. Therefore, to find $x \times y$, simply take the antilogarithm of the sum of the logs of x and y.

Another useful identity is obtained by replacing y in identity (11) by a second x. Then,

$$\log_{10}(x \times x) = \log_{10} x + \log_{10} x$$
$$= 2 \log_{10} x,$$

or

$$\log_{10} x^2 = 2 \log_{10} x.$$

In general,

$$\log_{10} x^n = n \log_{10} x. \tag{12}$$

This shows why logarithms were so useful in the days before calculators: they converted the relatively complicated operation of multiplication into the easier operation of addition. Correspondingly, calculations involving division could be reduced to the simpler operation of subtraction.

Activity 37   *Without your calculator*

As you may recall, Queen Calcula gave you one gold piece on the first day of the month, two on the second, four on the third, eight on the fourth, and so on, ending on the last day of a 31-day month. Given that $\log_{10} 2$ is about 0.3, calculate $\log_{10} 2^{30}$ *without* using your calculator, and hence estimate how many gold pieces the queen handed over on day 31.

Even though their use as a calculating tool has been superceded by the calculator, logarithms remain important for other reasons. In particular, logarithms are used implicitly in the definition of certain special measurement scales. One such scale that you may have come across is the Richter scale for earthquakes. This scale is rather unusual in that it works multiplicatively, not additively as most familiar scales of measurement do. An earthquake that measures, say, 6 on the Richter scale is 10 times more powerful than one that measures 5. The reason is that a Richter measure of 6 actually represents 10^6, while a measure of 5 represents 10^5. In the Richter system, going up 1 unit on the scale corresponds to multiplying the strength of the earthquake by a factor of 10. In other words, the Richter scale measure of an earthquake is given by the logarithm, to base 10, of its strength.

There is another well-known measurement scale that works in a similar way to the Richter scale: the decibel system for measuring the intensity of sound. This is slightly more complicated, because the fundamental unit is not actually the decibel but the bel—a decibel is simply one tenth of a bel. A sound level of 8 bels is 10 times stronger than one of 7 bels. To specify a sound level as so many bels is to give the *logarithm* of the intensity of the sound. It is worth bearing this in mind: for instance, if the noise level in your workplace is 'only' going up from 70 decibels (7 bels) to 80 decibels (8 bels), do not be taken in—the noise intensity is actually going to be 10 times greater than it was before.

4.4 Continuous exponential growth

Earlier in this unit you dealt with processes that occur in distinct stages: one day at a time, in the case of Queen Calcula's bounty; one year at a time, in the case of the growth of trees; one generation at a time, in the case of human ancestors. However, some processes vary continuously. Take, for example, the density and the pressure of the atmosphere, which decrease with increasing height above sea level: the average density and pressure are exponentially decreasing functions of height, but in neither case could you sensibly use the idea of 'a generation' in the analysis of how these quantities vary.

Another example of a continuously varying process is the cooling of a cup of hot coffee when it is left undisturbed in a moderate draught. The temperature of the coffee exhibits an exponential decay. Both the variables—temperature and time—are varying continuously even though, in practice, they are likely to be measured at distinct intervals.

In the case of Queen Calcula, both variables (the day of the month and the number of gold pieces) can only take values that are whole numbers. They do not vary continuously; instead they change from one whole number to another. Recall from *Unit 4* that the word used to describe models which show this stepping characteristic is 'discrete'. Thus Queen Calcula's generosity gives rise to a discrete exponential model, whereas the decrease in the temperature of a cup of coffee leads to a continuous exponential model.

Even though many of the models in this unit (for instance, for chain letters and for ancestors) are discrete, they have been analysed as though they were continuous, by using functions of the form $y = ab^x$. When a continuous function is used to model a discrete situation, you simply round to the nearest whole number or an appropriate number of decimal places.

In Section 5, you will meet exponential models that fall into an intermediate category between the complete discreteness of the problem of Queen Calcula's bounty, and the continuity of cooling coffee. This intermediate category arises when the changing population is very large (such as the number of atoms in a radioactive sample or the population of a country). The process can then usually be treated as if it were a continuous one.

In practice, all exponential models can be treated in the same way until the vital final step of interpreting any calculated results. At that point, the application of common sense should prevent you making such statements as 'Queen Calcula paid me 100 gold pieces on day 6.64'!

Before moving on to Section 5 where you will be applying what you know about exponentials, you may find it useful to pause and review what you have done so far in Sections 1–4.

Activity 38 *Taking an overview*

Think about what you have learned about exponentials. Use the learning outcomes at the end of each section to help you to identify important aspects of exponentials. Check that you can:

◇ explain what is meant by 'exponential growth';

◇ describe the characteristic features of the graph of an exponential function;

◇ understand the derivation of the formula for the sum of an exponential sequence and use it where appropriate;

◇ state the relationship between logarithms and exponentials;

◇ understand that exponential functions can be used to represent situations where the variables are either discrete or continuous.

Outcomes

After studying this section, you should be able to:

◇   understand and use the rules for calculating with exponentials (Activities 29 and 30);

◇   explain what is meant by b^x when x is zero; when x is a negative number; when x is a fraction (Activity 31);

◇   understand and use the relationship between logarithms and exponentials (Activities 32–37);

◇   understand that exponential functions can be used to represent situations where the variables are either discrete or continuous (Activity 38).

5 Exponentials in action

Aims   This section aims to help you apply what you have learned about exponentials to some important real-life problems.   ◇

As you work through this section, keep your own notes on the concepts introduced.

5.1 Doubling time

Consider a simple model of an insect population that increases in size by 10% each week, with a starting population size of 200 in week 0. The population size grows according to the exponential function $y = 200 \times 1.1^x$, where y is the population size and x is the week number. Here the population is discrete, but it is quite reasonable to think of x as continuous and not restricted to the values 0, 1, 2, 3, ... Although the population may only be counted, say, once per week, it is actually growing continuously. Therefore, it makes sense to consider questions such as: what was the population level when $x = 1.25$, or even $x = 0.3$? It would not be sensible, however, to say that the population was, for example, 300.5 insects.

▶   How long will it take for the population to reach 400 (that is, to double in size)?

The time it takes for a population to double is referred to as the *doubling time*, and it is a useful measure of how quickly a population is growing. It depends on the growth rate, but it presents the information about the growth of the population in a more accessible way than the growth rate itself does. Consequently, doubling times are often used instead of growth factors to represent the rates of growth of populations.

To find the doubling time for the insect population considered above, you need to solve the equation

$$200 \times 1.1^x = 400.$$

Divide both sides by 200 to get

$$1.1^x = 2.$$

This equation shows clearly that the doubling time depends *only* on the growth rate, and not at all on the initial population size. In other words, no matter what size of population you started with, it would take the same time to double.

The equation $1.1^x = 2$ can be solved graphically, by using the calculator to plot $y = 1.1^x$ and seeing where the graph crosses the line $y = 2$. But there is another way of solving it, which leads to a general formula for calculating doubling times.

The equation $1.1^x = 2$ is similar in form to the equation $10^x = 5$, which was solved at the outset of Section 4.3. There the solution was $x = \log_{10} 5$. So it should not come as a surprise to learn that the solution to the doubling-time problem also involves logarithms, but in a slightly different way.

The logarithm to base 10 of 1.1 is 0.0414, and the logarithm of 2 is 0.3010 (both to 4 decimal places). Hence $1.1 = 10^{0.0414}$ and $2 = 10^{0.3010}$.

Then the equation $1.1^x = 2$ may be rewritten as

$$(10^{0.0414})^x = 10^{0.3010}.$$

Applying the rule for raising an exponential to a power gives

$$10^{0.0414 \times x} = 10^{0.3010}.$$

The only way the two expressions above can be equal is for the exponents themselves to be equal. Therefore, it must be true that

$$0.0414 \times x = 0.3010,$$

and so

$$x = \frac{0.3010}{0.0414} = 7.27 \text{ (to 2 d.p.).}$$

Thus the doubling time of the insect population is about $7\frac{1}{4}$ weeks.

All these calculations could have been carried out to a much higher degree of accuracy, but the quoted number of decimal places is more than adequate for present purposes.

The final expression for the doubling time, x, of the insect population could also be written as

$$x = \frac{\log_{10} 2}{\log_{10} 1.1},$$

which is the solution of the equation $1.1^x = 2$.

This expression can be entered directly on your calculator to evaluate the doubling time.

Activity 39   Effects of changing the growth rate

If the growth rate of the insect population is reduced from 10% to 5%, the growth factor changes from 1.1 to 1.05. If the growth rate is increased from 10% to 20%, the growth factor increases to 1.2.

Calculate the doubling times of the insect population in the cases of 5% and 20% growth.

The same method as used for the insect population problem can be used to derive a formula for the doubling time of any population that is growing exponentially.

Suppose that the growth factor of the population is b, so the exponential function that gives the population size is $y = ab^x$, where a is the initial size of the population.

▶   What is the doubling time, d? Alternatively, what value of x gives a population size of $2a$?

To find the value of d, it is necessary to solve the following equation for d:

$$ab^d = 2a,$$

or, because a cancels out,

$$b^d = 2. \tag{13}$$

Notice that the doubling time, d, is independent of the initial population size. Also notice that the growth factor, b, must be bigger than 1, or the population will never actually double in size.

The method used earlier to solve an equation of the type $b^d = 2$ was to transform the equation into one involving powers of 10, by using logarithms. The logarithm of 2 is the power to which 10 must be raised to give 2; thus $10^{\log_{10} 2} = 2$. Likewise, $10^{\log_{10} b} = b$. So equation (13) can be rewritten as

$$(10^{\log_{10} b})^d = 10^{\log_{10} 2},$$

or

$$10^{d \times \log_{10} b} = 10^{\log_{10} 2}.$$

It follows that d must satisfy the equation

$$d \times \log_{10} b = \log_{10} 2.$$

Therefore,

$$d = \frac{\log_{10} 2}{\log_{10} b}. \tag{14}$$

This formula can be used to calculate the doubling time of any population that is growing exponentially and for which the growth factor b is known.

Activity 40   *Inflation and the pound in your pocket*

Due to inflation, the same goods usually cost more each year. You know from *Unit 2* that the percentage increase in the Retail Prices Index (RPI) over a year is the annual rate of inflation. Although the rate of inflation can vary a lot from year to year, it is useful to assume that the rate stays constant for a period of time, and ask how long it will take for the RPI to double. The doubling time for the RPI could equally well be described as the time it takes for the pound (£) to halve in value.

During the last 30 years or so (at the time of writing), the rate of inflation in the UK has fluctuated between a low of about 2% and a high of about 25%. How long would it take the RPI to double (or the pound to halve in value) if inflation were to stay constant at each of these extreme rates?

Activity 41   *Working backwards*

Sometimes you know what the doubling time is for a population that is growing exponentially, but you want to know the exponential growth rate.

The relation between the doubling time, d, and the growth rate, b, is given by formula (14) on the preceding page. Change the subject of that formula to express the growth rate in terms of the doubling time.

It was shown earlier in this section that the doubling time for an insect population with a weekly growth rate of 10% is about 7.27 weeks. This means that if you start with 200 insects, the population size will have increased to 400 in 7.27 weeks.

▶   How much longer would you have to wait for the population to double again, to 800?

Formula (14) does not depend on the size of the intial population so, again, the answer is 7.27 weeks. Therefore, once you have calculated the time it takes for the initial population to double, you know how long it will take for that population to double at other times (providing the growth rate is constant).

In general, if a population that is growing exponentially has a doubling time of d years, then in *any* period of d years, the population will double. For example, in d years from now, the population size will be twice what it is at present, and in $2d$ years from now, it will be 4 times what it is at present. It is this universal property of the doubling time that makes it so useful.

5.2 Half-life

The concept of a doubling time makes sense only for a population that is *growing* exponentially. However, there is a related concept for an exponentially *decaying* population—the time it takes for the population to *halve* in size. This is known as the *half-life*. You may already have met the half-life in the context of radioactive decay; for instance, 'the half-life of radioactive carbon is approximately 5730 years'.

The significance of this is shown in the following account:

> A church in Florence contains a terracotta cherub by the fifteenth-century Florentine sculptor, Donatello. A large crack in the cherub had been repaired using resin glue. Scientists decided to subject the glue to carbon dating so that they could make an estimate of when the repair was carried out.

> The results showed that the glue had been applied between AD 1398 and AD 1439—during the lifetime of the sculptor. Possibly the cherub cracked in the firing kiln, and Donatello repaired it himself.

Carbon dating relies on the fact that carbon is a chemical element that is found abundantly in the tissues of plants and all living things, and in the atmosphere (in the form of carbon dioxide). Most carbon that occurs naturally is non-radioactive; it is called carbon 12, or C^{12}. But there is a form of carbon that is very slightly radioactive; it is called radiocarbon, carbon 14 or C^{14}. The nucleus of an atom of C^{14} is unstable: at any moment it can spontaneously eject a β-particle and, when it has done so, it is no longer a carbon atom but an atom of nitrogen. Chemically, C^{14} is the same as ordinary carbon: it forms the same chemical compounds, and behaves in the same way in chemical reactions. The difference between the two forms of carbon lies only in the composition of their atomic nuclei.

Do not worry if the scientific detail of carbon dating is unfamiliar to you. Concentrate on the use of exponential ideas and the notion of half-life.

Such forms of an element are known as *isotopes* of the element concerned.

Suppose you had a lump of radiocarbon, consisting of a very large number of atoms. These atoms will spontaneously change into nitrogen atoms, one by one, but it is impossible to tell which radiocarbon atom will change when (this is what is meant by describing the decay as 'spontaneous'). However, half of the atoms will have turned into nitrogen in 5730 years, because the half-life of radiocarbon is 5730 years.

It has been found that the number of atoms of C^{14} in a sample decreases exponentially over time. Let y be the number of C^{14} atoms, t the time measured in half-lives, x the same time measured in years, and a the initial number of C^{14} atoms. Notionally, the dependent variable, the number of atoms, should be regarded as taking discrete values but practically, when dealing with such enormous numbers of atoms, individual atoms count for scarcely anything. So the variables may be treated as continuous.

Because t half-lives are the same as x years,

$$t \times 5730 = x \quad \text{and} \quad t = \frac{x}{5730}.$$

Since the growth factor $b = \frac{1}{2}$, the formula giving the number of C^{14} atoms at any time is

Recall formula (1) on page 18.

$$y = a \times \left(\tfrac{1}{2}\right)^t = a \times \left(\tfrac{1}{2}\right)^{x/5730} = a \times \left[\left(\tfrac{1}{2}\right)^{1/5730}\right]^x.$$

But $\left(\tfrac{1}{2}\right)^{1/5730}$ is $0.999\,879$ (to 6 decimal places). So

$$y = a \times (0.999\,879)^x.$$

▶ How could scientists use the property of the radioactive decay of radiocarbon to date objects such as wood, resin glue, and so on?

The procedure is based on the fact that C^{14} is present in the carbon dioxide in the atmosphere in small quantities, as a result of the bombardment of the atmosphere by cosmic rays. The creation of new C^{14} in the atmosphere by cosmic rays pretty well balances out the decay of the C^{14} already there, so the proportion of radiocarbon in atmospheric carbon stays constant. When a plant is growing it takes in C^{14} in the same proportion relative to ordinary C^{12} as the corresponding proportion that exists in the atmosphere. After the plant dies it no longer absorbs any C^{14}, and the C^{14} that it already contains undergoes radioactive decay.

The amount of C^{14} in dead plant material is, therefore, continually declining. The decay is slow since the half-life of C^{14} is 5730 years. Nevertheless, over a period of a few hundred years there is a measurable reduction in the C^{14} level; this can be determined by measuring the corresponding reduction in radioactivity. In this way, it is possible to calculate how long ago a sample of wood or linen or other plant-based material stopped growing, and so date the artefact from which the sample was taken.

In the case of the Donatello cherub, the glue that provided the sample contained resin, which comes from pine trees. Because the half-life of radiocarbon is comparatively long, dating samples like this one which are not terribly old requires great precision. It is possible to estimate the reduction in the amount of radiocarbon in the resin by calculating $(0.999\,879)^x$, where x is the age of the sample in years. In this instance, if the crack in the cherub had been repaired during the lifetime of the sculptor, that is, about 600 years ago, then x would be 600. Now $(0.999\,879)^{600} \simeq 0.93$. So if the suggestion about the date of the repair is correct, then the amount of C^{14} in the sample should have dropped by about 7%, compared with the amount there originally. However, if the crack had been repaired only 500 years ago, then $(0.999\,879)^{500} \simeq 0.94$, and the amount of C^{14} in the sample would have dropped by about 6%. This gives an idea of how small the effect is that has to be measured when radiocarbon dating is used to find the age of something made in historical times.

Here, then, is a summary of the procedure for carbon dating.

The first step in dating a sample like the glue from the Donatello statue is to extract the carbon by burning the sample to convert it to graphite. Next, the amount of radioactivity associated with this carbon has to be measured. The result is then compared with how much radioactivity is produced by the same quantity of atmospheric carbon. This comparison shows how the proportion of C^{14} in the sample has decreased since its carbon content was fixed when, for example, the resin in the glue was extracted from the tree. Finally, the exponential decay formula, based on the known half-life of C^{14}, can be used to calculate the age of the sample.

Example 8 *Carbon dating in practice*

Suppose an ancient wooden bowl was discovered and tests showed that the ratio of C^{14} to C^{12} was 0.84 times the ratio of C^{14} to C^{12} in the atmosphere. How old is the bowl?

Solution

Let x be the age of the bowl in years. Then $(0.999\,879)^x = 0.84$. Taking logarithms of both sides:

$$x \log_{10} 0.999\,879 = \log_{10} 0.84,$$

so

$$x = \frac{\log_{10} 0.84}{\log_{10} 0.999\,879}$$
$$\simeq 1440.$$

Therefore the bowl is roughly 1440 years old.

Activity 42 *More carbon dating*

A measurement of the radioactivity of a sample of wood from an archeological dig shows that the ratio of C^{14} to C^{12} in the sample is 0.71 times the ratio of C^{14} to C^{12} in the atmosphere. Estimate the date of the site that is being excavated.

Activity 43   *Carbon dating: assumptions*

The carbon dating method depends for its reliability on an assumption about how the ratio of C^{14} to C^{12} in the atmosphere varies. What is that assumption?

Many other elements have isotopes that are capable of undergoing radioactive decay. In any radioactive decay, an unstable atomic nucleus spontaneously disintegrates by emitting an α- or a β-particle (possibly accompanied by a γ-ray). As a result, the atom changes into another element. The new atom may again be an unstable isotope.

In general, the number of atoms, y, of any radioactive isotope will decrease exponentially according to the formula

$$y = a \times \left(\tfrac{1}{2}\right)^{t}$$

or, since $\tfrac{1}{2} = 2^{-1}$,

$$y = a \times (2)^{-t},$$

where a is the initial number of atoms and t is the time measured in half-lives. This alternative form of the formula draws attention to the fact that a half-life is quite similar to a doubling time, except that the half-life specifies *how long ago* there were twice as many atoms of the radioactive isotope in the sample as there are now.

5.3 APR

What happens if the time scale of an exponential change is altered in some way (from years to months, for instance)?

A relevant financial example involves changing the frequency with which interest is compounded. If you have a credit card, you are charged interest on any debts that you owe to the credit card company. Statements are presented monthly, and if you fail to pay off your debt at the end of the month, then you will be charged interest on the sum outstanding on the next statement. So interest is charged on a monthly basis. However, the credit card company is required by law to publish the equivalent *annual percentage rate* (APR). The idea of the APR is to present the interest rate in a form that enables you to make a sensible comparison between using a credit card as a form of loan, and other forms of loan, such as bank loans and mortgages, for which interest may be charged on a different basis.

The APR is calculated on the following basis. It is supposed that you have a credit card debt of £100 at the beginning of the year, and that you make no transactions and pay nothing off in the year. Interest will be charged monthly and added to the sum outstanding. Then, the *total* amount of interest owing at the end of the year gives the APR for that account.

▶ What is the APR if the monthly interest rate is 2%?

If the monthly interest rate is 2%, and you owe £100 initially, then after one month you will owe £102. After two months you will owe this amount plus 2% of £102; that is, $1.02^2 \times £100$. This process continues on the familiar exponential pattern, so after one year you will owe $1.02^{12} \times £100$.

Now $1.02^{12} \simeq 1.268\,241\,795$; therefore, the sum you owe at the end of the year is £126.82 (to the nearest penny). The total amount of interest owing is £26.82, and hence the APR is 26.82%.

Note that, because the calculation of the APR involves an exponential, the APR is larger than the monthly rate multiplied by 12. This, indeed, is one of the reasons why credit card companies are required to quote the interest rate in the form of an APR. People are easily misled by small rates compounded over short times into thinking that they are getting a bargain, when on an annual basis they are paying much more.

▶ What is the monthly interest rate if the APR is 24%?

This is the converse question to the one above. When the APR is 24%, the total interest owed on £100 at the end of the year is £24. This means that the total sum owed (original debt plus interest) is £124. If x% is the unknown monthly interest rate, then

$$\left(1 + \frac{x}{100}\right)^{12} \times 100 = 124,$$

or

$$\left(1 + \frac{x}{100}\right)^{12} = 1.24.$$

Taking the twelfth root of both sides gives

$$1 + \frac{x}{100} = \sqrt[12]{1.24}$$

$$\frac{x}{100} = \sqrt[12]{1.24} - 1,$$

hence

$$x = 100\left(\sqrt[12]{1.24} - 1\right)$$
$$= 1.81 \text{ (to 2 d.p.)}.$$

The monthly interest rate corresponding to an APR of 24% is, therefore, 1.81% (to 2 decimal places).

Activity 44 *APR revisited*

Find (i) the APR for a monthly interest rate of 1%, and (ii) the monthly interest rate for an APR of 12%.

Activity 45   More frequent interest

The concept and definition of the APR are not just restricted to monthly payments and, indeed, daily rates are now quite commonly quoted. Find the APR for an interest rate of 0.07% compounded daily (that is, 365 times a year).

Note that in Activity 45, time is being treated as a discrete variable, but, as computing power increases, the time intervals quoted by lending institutions are becoming shorter and shorter, and the variable becomes a closer approximation to a continuous one.

Activity 46   More on interest

Suppose that you need a loan of £1000, and want to pay as little interest as possible. You are offered a choice of three fixed-interest loans:

(a) an interest rate of 6.75%, compounded annually;

(b) an interest rate of 3.25%, compounded every six months;

(c) an interest rate of 0.5%, compounded monthly.

Calculate the equivalent APRs and decide which option you should choose.

Activity 47   Interest again

Suppose that you are offered the choice of two savings accounts paying interest at fixed rates for one year:

(a) an interest rate of 0.03%, compounded daily;

(b) an interest rate of 0.2%, compounded weekly.

Which will give you the better return on your savings?

When reading an advertisement offering a loan at a certain APR, with interest to be repaid monthly, it is natural to make a quick estimate of the monthly interest rate by dividing the APR by 12. But this leads to an overestimate of the rate.

As a very simple example of this, consider the case where the APR is 100%. This means that the sum owing doubles in a year, so if the original debt were £1, the sum owing after 1 year would be £2. Alternatively, if the interest were compounded twice at an interest rate of half the APR, that is at 50%, then the total owing at the end of the year would be £1.5^2 = £2.25. The effect of the approximation of simply dividing the APR by the number of times that the interest is compounded is to produce an overestimate of the total sum owing by £0.25 in this case.

If the number of times that the interest is compounded is increased, say to 4, then the error in the approximation gets worse: if the quarterly interest rate is taken to be 25%, the total owing would be £1.25^4 = £2.44. Further increasing the number of times that the interest is compounded increases the error still more. In the next activity you can investigate just how bad the error can be.

Activity 48   *Showing too much interest*

Use your calculator to complete the following table based on a loan of £1 over a period of 1 year with an APR of 100%.

Number of times per year interest is compounded	Total sum owing at end of one year /£
1	2
2	2.25
4	2.44
12	2.61
24	2.66
52	
365	
500	
1000	
10000	

In fact, what you have been computing in Activity 48 is the expression

$$\left(1 + \frac{1}{n}\right)^n$$

for increasing values of n, where n is the number of times that interest is compounded (the number in the left-hand column of the table in Activity 48). As n gets larger and larger, the numbers in the right-hand column of the table tend to the value 2.718 281 828 5.... This is the constant e, which you met in Section 3.2.

5.4 Populations of people

As part of your work in this section, you will have to use the calculator to find the best exponential fit to some given data. The technique for doing this is explained in the *Calculator Book*.

Now study Section 12.4 of Chapter 12 in the Calculator Book

The growth of a real population (that is, a population of people) is a good example of a discrete process that has to be treated as a continuous one, even though individual people cannot be subdivided.

For example, consider the population of the USA, which has grown as in Figure 6. From the shape of this graph, it looks as if this population may exhibit exponential growth.

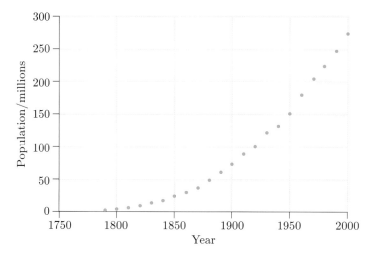

Figure 6   Population of the United States, 1790–2000. Source: Population Estimates Program, Population Division, US Census Bureau.

To model such a population, it is a reasonable assumption that, for any fixed short period of time (for instance, a day—but it could be an hour, or a week—choose whichever you fancy), the increase in the population is proportional to the size of the population. The number of babies born depends, after all, on how many women there are to have them: the greater the size of the population, then (other things being equal) the more babies will be born. Likewise, a fixed proportion of the population dies each day. Assume that this proportion—the death rate—as well as the birth rate are constant over time. To keep things simple, ignore the complication of migration. The population size over the course of a day can then be modelled as follows:

population at end of day = population at beginning of day
+ births − deaths.

If $P(n)$ is the size of the population on the nth day, B is the birth rate (the number of live births per head of the population per day) and D is the death rate (the number of deaths per head of the population per day), then substituting these symbols into the word formula gives

$$P(n+1) = P(n) + BP(n) - DP(n)$$
$$= (1 + B - D)P(n).$$

Thus

$$P(n+1) = \text{some constant} \times P(n).$$

This resembles the familiar formula for exponential growth, given as formula (3) on page 19. So the model predicts exponential population growth. Whether or not the population actually grows at all depends on whether or not $B > D$.

If you have data on a population size, you can try to fit an exponential function to them to see how well the model predictions fit, as the next activity demonstrates.

Activity 49   *World population growth*

Table 6 gives estimated data for the world's population.

Table 6

Year (x)	Population (y)
1850	1.094×10^9
1900	1.500×10^9
1925	1.907×10^9
1930	2.070×10^9
1950	2.500×10^9
1960	3.019×10^9
1990	5.292×10^9
2000	6.080×10^9

Source: The data for 1930, 1960 and 1990 come from the *United Nations Demographic Yearbook 1990*; the remaining data are from *The Future Growth of World Population* (UN, 1958) and the US Census Bureau, International Data Base 2000.

(a) Use your calculator to produce a scatterplot of the data in Table 6.

(b) Use the exponential regression facilities on your calculator to find the values of a and b for which the graph of $y = ab^x$ best fits the data. With these values of a and b, superimpose the graph of $y = ab^x$ on your scatterplot so that you can see how well the exponential model fits the data.

In the more recent years for which data are available, it seems that the world's population has been growing more quickly than the historic trend would have suggested.

▶   Can you offer any explanations for this discrepancy?

Possible explanations might include improved preventive medicine and medical treatments, better nutrition, and reduction in death rates.

One of the best ways of understanding the consequences of the continuing growth in the world's population is to work out how long it will take for the population to double; in other words, to calculate the doubling time. This is a statistic that is often quoted in discussions of world population growth.

Activity 50   *Population doubling time*

Use the best exponential fit to the data from Activity 49 to find the doubling time for the world's population.

In the 40 years from 1960 to 2000, the world's population has increased more than predicted by the model that is based on the historical data. In fact, the ratio of the population in 2000 to the population in 1960 is $6.080 \div 3.019 = 2.014$. This shows a doubling time of about 40 years, compared with a doubling time of about 59 years predicted by the model. So, the population of the world at the end of the twentieth century was actually growing more quickly than the exponential model predicted.

Outcomes

After studying this section, you should be able to:

◇   explain the terms 'doubling time' and 'half-life', and calculate these quantities for an exponentially growing or decaying population (Activities 39–42);

◇   explain what an APR is, and calculate it when interest is charged for some period other than a year (Activities 44–48);

◇   fit an exponential regression model to given data and interpret the results (Activities 49 and 50).

Unit summary and outcomes

Just as *Unit 10* looked at the family of linear functions and *Unit 11* at the family of quadratic functions, this unit has focused on a third family, that of exponential functions (and their related inverse functions, the *logarithms*). The applications investigated have been widespread and have included ancestors, chain letters, investments, populations and radioactivity.

Outcomes

You should now be able to:

◇ explain what is meant by 'exponential growth', and recognize examples of such growth (Activities 1, 2 and 5);

◇ write down and use the general formulas for the size of a population that is growing exponentially (Activities 3–9);

◇ interpret exponential models (Activity 10);

◇ decide, in a given situation involving exponential change, when it is appropriate to calculate the sum over several generations (Activities 11, 12, 14 and 15);

◇ understand the derivation and use of the formulas for the cumulative sum (Activities 11–15);

◇ convert exponential functions into the general form $y = ab^x + c$ (Activity 16);

◇ describe the characteristic features of the graphs of exponential functions (Activities 17–20);

◇ describe how the values of the constants a and b affect the graph of $y = ab^x$ (Activities 21–25);

◇ describe the graph of $y = ab^x + c$ and draw conclusions about the long-term behaviour of a population when the growth factor is between 0 and 1 (Activities 26 and 27);

◇ describe the graph of *the* exponential function $y = e^x$ and its gradient function (Activity 28);

◇ understand and use the rules for calculating with exponentials (Activities 29 and 30);

◇ explain what is meant by b^x when x is zero; when x is a negative number; when x is a fraction (Activity 31);

◇ understand and use the relationship between logarithms and exponentials (Activities 32–37);

◇ understand that exponential functions can be used to represent situations where the variables are either discrete or continuous (Activity 38);

◇   explain the terms 'doubling time' and 'half-life', and calculate these quantities for an exponentially growing or decaying population (Activities 39–42);

◇   explain what an APR is, and calculate it when interest is charged for some period other than a year (Activities 44–48);

◇   fit an exponential regression model to given data and interpret the results (Activities 49 and 50).

Comments on Activities

Activity 1

(a) Exponential.

The population consists of the bacterial cells.

The growth factor is 2.

(b) Not exponential because the distances travelled in successive time intervals go up by amounts related to the *square* of the time and are therefore not multiplied by a fixed factor.

(c) Traditionally, the point of this riddle is that only 'I' was *going to* St Ives but met the others *coming from* there.

Exponential.

The population consists of the things coming from St Ives.

The growth factor is 7.

(d) Not exponential because the record times have decreased by the same *amounts* over two equal time intervals. There is no multiplicative growth factor at work here.

Activity 2

The following example, pointing out the misuse of the word 'exponentially', was found on the web:

'...What do I read as the first line of the letter introducing Wilmott Research Report 6? "Registrants to Wilmott.com continue to sign up in exponentially large numbers." There are no exponentially large numbers. Numbers are fixed. That's why we call them "numbers" and not "variables". Variables can increase exponentially, ...'

Aaron Brown, Tuesday, 16 October 2001

http://www.wilmott.com/310/messageview.cfm?catid=15&threadid=204

Activity 3

By analogy with the method used in calculating the numbers of ancestors, the numbers of letters sent at the specified stages are:

$$5^3; \ 5^{17}; \ 5^n.$$

Activity 4

The number of letters sent at the tenth stage is $P = 5^{10} = 9\,765\,625$.

Activity 5

By analogy with Example 2, the balance after 5 years, $P(5) = 500 \times 1.04^5 = £608.33$ (to the nearest penny).

With simple interest, the balance would have increased by $5 \times 4\%$ over the 5-year period to give

$$£500 + £500 \times 5 \times 0.04 = £600.$$

Activity 6

By analogy with Example 2, the balance after n years, $P(n) = £A \times 1.03^n$.

Activity 7

If the rock star invests $£A$, then after 18 years she will have $£A \times 1.05^{18}$. So

$$A \times 1.05^{18} = 1\,000\,000.$$

Hence

$$A = 1\,000\,000 \div 1.05^{18} = 415\,520.65.$$

This means that the sum that the rock star must invest initially is £415 520.65 (to the nearest penny).

Activity 8

There will be 17×3^n simplified greenfly on the bush n days later (that is, n days after 1 May because the newborn greenfly produce offspring only on the day *after* their birth).

On 11 May, $n = 10$. So there will be $17 \times 3^{10} \simeq 1$ million greenfly on the bush.

Activity 9

(a) Following on from Example 3,
$$P(n) = 3 \times \left(\tfrac{4}{3}\right)^n.$$

(b) When the sides of the original triangle are all $\tfrac{1}{3}$, the perimeter of that triangle is 1. So
$$P(n) = 1 \times \left(\tfrac{4}{3}\right)^n = \left(\tfrac{4}{3}\right)^n.$$

If the perimeter of the original triangle is 5, then
$$P(n) = 5 \times \left(\tfrac{4}{3}\right)^n.$$

(c) If the perimeter of the original triangle is l, the perimeter of the figure at the nth stage
$$P(n) = l \times \left(\tfrac{4}{3}\right)^n.$$

Activity 10

(a) The term $kP(n)$ corresponds (in the reverse order) to the (number of adult females in breeding population in year n) × (proportion of adults surviving to following year).

So far as the other given term is concerned, the first bracket, (number of young females produced in year n), is obtained by taking the number of breeding females, $P(n)$, and multiplying by half the average number of young, m. This gives $P(n) \times \tfrac{1}{2}m$. Then, this is multiplied by the second bracket, (proportion of young surviving to following year), which is l, to give
$$P(n) \times \tfrac{1}{2}m \times l.$$

Hence the word model translates into
$$P(n+1) = kP(n) + \tfrac{1}{2}mlP(n).$$

(b) Substituting the values 0.86, 0.60 and 4 for k, l and m, respectively, into the equation $b = k + \tfrac{1}{2}ml$ gives $b = 2.06$.

(c) Rounding to 3 significant figures, the population sizes in successive years, according to the model, are

1961	1000×2.06	$= 2060,$
1962	$1000 \times (2.06)^2$	$= 4240,$
1963	$1000 \times (2.06)^3$	$= 8740,$
1964	$1000 \times (2.06)^4$	$= 18\,000.$

These figures are in reasonable agreement with the data shown in Figure 5.

(d) The population size will increase sharply through the breeding season, as young are produced in great numbers; but it will then gradually decline until the beginning of the next breeding season as old age and winter take their toll. It is therefore important to specify carefully when the population is measured, in order to compare like with like from one year to the next, and to ensure that what is predicted by the model corresponds to what is counted by the ornithologists in the field.

(e) As a species new to the British Isles, the collared dove probably found a niche for itself: that is, it found little competition for its food and few predators. Moreover, it was a protected species initially and could not be shot.

The population could not continue to grow exponentially indefinitely because, if it did, it would get arbitrarily large. Eventually growth had to slow down, as competition for food and living space took over. In fact, after 1965, growth rates reduced considerably though the population continued to increase to an estimated 400 000 in 1988–91.

Activity 11

From the formula $S(n) = 2^n - 1$, the total number of gold pieces handed over by the queen in February is $2^{28} - 1 = 268\,435\,455$ in ordinary years, and $2^{29} - 1 = 536\,870\,911$ in leap years.

Activity 12

In effect, there are five generations—man, wives, sacks, cats and kits—and the growth factor is 7, so putting $n = 5$ and $b = 7$ in formula (5) gives

$$S(5) = \frac{7^5 - 1}{7 - 1} = \frac{16\,806}{6} = 2801.$$

Activity 13

The formula to be rewritten is

$$S(n) = a(1 + b + b^2 + \cdots + b^{n-1}).$$

From formula (5), you know that

$$1 + b + b^2 + b^3 + \cdots + b^{n-1} = \frac{b^n - 1}{b - 1}.$$

So substituting into the initial formula gives

$$S(n) = a\left(\frac{b^n - 1}{b - 1}\right).$$

Another way of looking at this is to recognize that in the general case where the initial population size is a, each term is a times the corresponding term when the initial population size is 1. So the overall sum, $S(n)$, will be bigger by a factor of a. Therefore, by comparison with formula (5), the general sum will be

$$S(n) = a\left(\frac{b^n - 1}{b - 1}\right).$$

Activity 14

At the end of year 1,

$$D(1) = 1000 \times 1.04 - 100.$$

At the end of year 2,

$$\begin{aligned} D(2) &= (1000 \times 1.04 - 100) \times 1.04 - 100 \\ &= 1000 \times 1.04^2 - 100(1 + 1.04). \end{aligned}$$

This pattern continues in a similar way to that in Example 6. So, at the end of year n,

$$D(n) = 1000 \times 1.04^n - 100\left(\frac{1.04^n - 1}{1.04 - 1}\right).$$

Rearranging and simplifying gives

$$\begin{aligned} D(n) &= 1000 \times 1.04^n - 100\frac{(1.04^n - 1)}{0.04} \\ &= 1000 \times 1.04^n - 2500(1.04^n - 1) \\ &= 1000 \times 1.04^n - 2500 \times 1.04^n + 2500 \\ &= 2500 - 1500 \times 1.04^n. \end{aligned}$$

To find what you owe at the end of year 5, substitute $n = 5$ into the above formula:

$$\begin{aligned} D(5) &= 2500 - 1500 \times 1.04^5 \\ &= 675.02. \end{aligned}$$

Therefore you owe £675.02.

Activity 15

At the end of day 1, the initial $\frac{1}{3}$ ration of water in the soil becomes $1 + \frac{1}{2} \times \frac{1}{3}$. So

$$\begin{aligned} w(1) &= 1 + \tfrac{1}{2} \times \tfrac{1}{3}, \\ w(2) &= 1 + \tfrac{1}{2}\left(1 + \tfrac{1}{2} \times \tfrac{1}{3}\right) \\ &= 1 + \tfrac{1}{2} + \tfrac{1}{3} \times \left(\tfrac{1}{2}\right)^2, \\ w(3) &= 1 + \tfrac{1}{2}\left[1 + \tfrac{1}{2} + \tfrac{1}{3} \times \left(\tfrac{1}{2}\right)^2\right] \\ &= 1 + \tfrac{1}{2} + \left(\tfrac{1}{2}\right)^2 + \tfrac{1}{3} \times \left(\tfrac{1}{2}\right)^3, \\ w(4) &= 1 + \tfrac{1}{2} + \left(\tfrac{1}{2}\right)^2 + \left(\tfrac{1}{2}\right)^3 + \tfrac{1}{3} \times \left(\tfrac{1}{2}\right)^4, \end{aligned}$$

and so on.

Therefore, the formula for day n is

$$w(n) = 1 + \tfrac{1}{2} + \left(\tfrac{1}{2}\right)^2 + \cdots + \left(\tfrac{1}{2}\right)^{n-1} + \tfrac{1}{3} \times \left(\tfrac{1}{2}\right)^n.$$

You can simplify this by using formula (7) for the terms up to $\left(\frac{1}{2}\right)^{n-1}$. This gives

$$\begin{aligned} w(n) &= \left[\frac{1 - \left(\tfrac{1}{2}\right)^n}{1 - \tfrac{1}{2}}\right] + \tfrac{1}{3} \times \left(\tfrac{1}{2}\right)^n \\ &= 2 - 2 \times \left(\tfrac{1}{2}\right)^n + \tfrac{1}{3} \times \left(\tfrac{1}{2}\right)^n \\ &= 2 - \tfrac{5}{3} \times \left(\tfrac{1}{2}\right)^n. \end{aligned}$$

Compare this with the expression $w(n) = 2 - 2 \times \left(\frac{1}{2}\right)^n$ for an initially dry soil. Once again, a steady state of 2 rations of water will be achieved within a couple of weeks—surprisingly the same steady state as when the farmer started with dry soil.

Activity 16

(a) $y = 3 \times \left(\frac{4}{3}\right)^x$;

$a = 3$, $b = \frac{4}{3}$, $c = 0$.

(b) $y = 5^x$;

$a = 1$, $b = 5$, $c = 0$.

(c) $y = 2^x - 1$;

$a = 1$, $b = 2$, $c = {}^-1$.

(d) $y = 6000 - 1000 \times 1.05^x$;

$a = {}^-1000$, $b = 1.05$, $c = 6000$.

Activity 17

(a) The graph of $y = 3^x$ lies entirely above the x-axis and passes through $(0, 1)$ on the y-axis. It goes shooting up with an ever-increasing gradient as you move to the right of the y-axis, but it hugs the x-axis more and more closely as you go to the left.

(b) By contrast, a straight line (with positive slope) always goes uphill to the right, but with no change of steepness. A parabola (the right way up) goes uphill with increasing gradient as you move to the right of its vertex, but it also goes uphill as you move to the left of its vertex.

Activity 18

(a) When $x = 0$, $y = 1$.

When $x = 1$, $y = 3$.

When $x = {}^-1$, $y = \frac{1}{3}$.

(b) The value of 3^x gets rapidly larger and larger as x increases from 0.

It decreases, getting closer and closer to 0 as x becomes more and more negative.

Activity 19

(a) Both graphs pass through $(0, 1)$. Also they both always have a positive gradient.

The straight line goes below the x-axis; the exponential never does.

(b) Both graphs lie above the x-axis.

The parabola is symmetric around the y-axis; the exponential is not.

Activity 20

(a) As x tends to $+\infty$, $2x + 1$ tends to $+\infty$.
As x tends to $^-\infty$, $2x + 1$ tends to $^-\infty$.

(b) As x tends to $+\infty$, $2x^2 + 1$ tends to $+\infty$.
As x tends to $^-\infty$, $2x^2 + 1$ tends to $+\infty$.

Activity 21

Your graphs should look like those below.

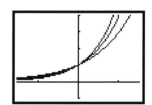

(a) When $x = 0$, then $y = 1$ in all cases.

When $x = 1$, then $y = 3$, 4 and 5, respectively.

(b) The graph of $y = 6^x$ would lie *above* the other three graphs when $x > 0$, but *below* them when $x < 0$.

The graph of $y = 2^x$ would lie *below* the other three graphs when $x > 0$, but *above* them when $x < 0$.

The graph of $y = 3.5^x$ would lie *between* $y = 3^x$ and $y = 4^x$.

(c) All graphs of the form $y = b^x$ have the same general shape for $b > 1$. The shape is characterized by the following features.

As x tends to $-\infty$, y tends to 0 ($y = 0$ is an asymptote). At $x = 0$, $y = 1$. Then the gradient increases and, as x tends to $+\infty$, y also tends to $+\infty$.

If $B > b$, then the graph of $y = B^x$ lies *above* that of $y = b^x$ when $x > 0$, but *below* it when $x < 0$. The graphs cross on the y-axis, at the point $(0, 1)$.

Activity 22

The graphs should resemble those below.

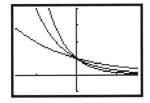

(a) When $x = 0$, then $y = 1$ in all cases.

When $x = 1$, then $y = 0.5$, 0.25 and 0.125, respectively.

When $x = {}^-1$, then $y = 2$, 4 and 8, respectively.

(b) All three graphs show exponential decay, and are characterized by the following features. As x tends to $-\infty$, y tends to $+\infty$. At $x = 0$, $y = 1$. As x tends to $+\infty$, y tends to 0 ($y = 0$ is an asymptote).

When $x < 0$, $y = 0.5^x$ lies below $y = 0.25^x$ which, in turn, lies below $y = 0.125^x$. The graphs are in the opposite order when $x > 0$.

The graph of $y = 0.6^x$ would lie below $y = 0.5^x$ when $x < 0$, and above it when $x > 0$.

The graph of $y = 0.2^x$ would lie between the graphs of $y = 0.25^x$ and $y = 0.125^x$.

The graph of $y = 0.025^x$ would lie below $y = 0.125^x$ when $x > 0$, and above it when $x < 0$.

(c) The general shape of such a graph corresponds to exponential decay.

As x tends to $+\infty$, b^x tends to 0.
As x tends to ${}^-\infty$, b^x tends to $+\infty$.

If $B > b$ and both are between 0 and 1, the two graphs $y = b^x$ and $y = B^x$ are similar in shape, but $y = B^x$ lies below $y = b^x$ when $x < 0$, and above it when $x > 0$.

(d) Graphs of both forms pass through $(0, 1)$. The graph of $y = B^x$ shows the classic exponential growth, whereas that of $y = b^x$ shows exponential decay. The shapes are roughly mirror images of each other. (They are the exact mirror images of each other when $B = 1/b$.)

Activity 23

(a) Because 1 raised to any power is still 1, it follows that the graph of $y = 1^x$ is the same as the graph of $y = 1$; that is, it is a straight line parallel to the x-axis and passes through the point $(0, 1)$.

(b) The point $(0, 1)$.

Activity 24

Your graphs should look like those below.

(a) When $x = 0$, then $y = 1$, 3, 4 and 0.3, respectively.

(b) The graphs are similar in shape, both to each other and to the general graph of $y = b^x$ for $b > 1$. Also, all graphs of the form $y = a \times 2^x$ cut the y-axis at $y = a$.

The graphs differ in that the graph of $y = 3 \times 2^x$ lies below that of $y = 4 \times 2^x$ for *all values* of x—the two graphs never cross.

Activity 25

(a) The graph of $y = {}^-2^x$ looks like that below (a y-range of $^-4$ to 1 has been used).

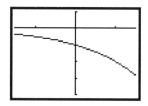

(b) The graph of $y = {}^-2^x$ is the mirror image (in the x-axis) of the graph of $y = 2^x$.

Activity 26

(a) The graph of $y = 2^x - 1$ is the same shape as that of $y = 2^x$, but it is shifted down by one unit.

(b) The graph of $y = 2^x + 1$ is the same shape as that of $y = 2^x$, but it is shifted up by one unit.

(c) The debt reduces to £0 when $y = 0$. This occurs when the graph crosses the x-axis—in this case, at $x \simeq 36.7$. So the debt reduces to zero in just under 37 years.

Activity 27

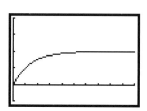

The graph of $y = 2 - 2 \times \left(\frac{1}{2}\right)^x$ climbs up to the line $y = 2$, and soon becomes indistinguishable from it. So, as x tends to $+\infty$, y tends to 2 ($y = 2$ is an asymptote).

Activity 28

Here are some sample values obtained by this process.

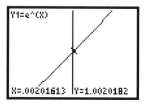

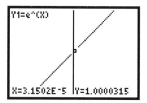

You should find similar values to these but your values will depend on your initial window settings and how far you zoom in.

Since x and y increase by similar amounts, the gradient must be very close to 1 at $(0, 1)$.

Activity 29

(a) You simply add together the number of zeros in each of the terms. This means that the answer will be 1 followed by 18 zeros.

(b) Ten billion millions $= 10 \times 10^9 \times 10^6 = 10^{16}$;

one hundred billion billions
$$= 100 \times 10^9 \times 10^9 = 10^{20};$$

one thousand million billion million billions
$$= 10^3 \times 10^6 \times 10^9 \times 10^6 \times 10^9 = 10^{33}.$$

Activity 30

(a) The mean distance from the Earth to the Sun is roughly $(1.5 \times 10^8) \times (3.9 \times 10^4)$
$$= 5.85 \times 10^{12} \text{ inches.}$$

(b) The mean distance from Pluto to the Sun is roughly $(5.9 \times 10^9) \times (3.9 \times 10^4)$
$$= 23.01 \times 10^{13} \text{ inches.}$$

To write this in scientific notation, make the first factor a number between 1 and 10. Then the answer becomes 2.301×10^{14} inches.

Activity 31

(a) Both answers are 1.6818 (to 4 d.p.) The two answers are the same because both the expressions $\sqrt[4]{2^3}$ and $\left(\sqrt[4]{2}\right)^3$ are equivalent to $2^{\frac{3}{4}}$.

(b) From your calculator you should find that $10^{0.75} = 5.623413252$. When this is raised to the power of 4, the answer is 1000, because $(10^{0.75})^4 = 10^{\frac{3}{4} \times 4} = 10^3 = 1000$.

(c)
$$8^{2/3} = \left(\sqrt[3]{8}\right)^2 = 2^2 = 4;$$
$$9^{3/2} = \left(\sqrt{9}\right)^3 = 3^3 = 27;$$
$$\sqrt[3]{2^6} = 2^{6/3} = 2^2 = 4;$$
$$\sqrt[4]{100^2} = 100^{2/4} = 100^{1/2} = \sqrt{100} = 10.$$

Activity 32

The graphs of $y = 10^x$ and $y = 5$ look like those below.

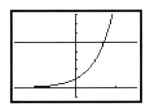

The x-coordinate of the point where the two graphs intersect (which is the solution of the equation $10^x = 5$) is 0.699 (to 3 d.p.). This is found using the trace and zoom facilities in the usual way.

Activity 33

$$10^{0.7} = 5.011\,872\,336;$$
$$10^{0.699} = 5.000\,345\,35;$$
$$10^{0.698\,97} = 4.999\,999\,95.$$

Activity 34

Your calculator should show that
$$\log_{10} 5 = 0.69897\ldots$$
$$= 0.699 \text{ (to 3 d.p.)}.$$

Activity 35

(a) $10^{2.34} = 218.776\,162\,4;$
$\log_{10}(218.776\,162\,4) = 2.34.$

(b) $\log_{10} 3.5 = 0.544\,068\,044\,4;$
$10^{0.5440680444} = 3.5.$

Activity 36

The graphs look like this, if drawn using the ZDecimal window settings.

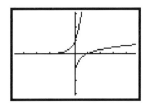

If you reflect one of these graphs in a line at $45°$ to the x-axis, you get the other graph.

Activity 37

The number of gold pieces handed over on day 31 was 2^{30}.

Now $\log_{10} 2^{30} = 30\log_{10} 2$, and since $\log_{10} 2 = 0.3$, it follows that
$$\log_{10} 2^{30} = 30 \times 0.3 = 9.$$
So
$$2^{30} \simeq 10^9.$$

Therefore you would have received roughly one billion gold pieces on the last day of the month.

Activity 38

In reviewing what you have done, make sure that the Handbook notes you have made are accurate, complete and easy to use.

Activity 39

When the growth rate is 5%,

$$\text{doubling time} = \frac{\log_{10} 2}{\log_{10} 1.05}$$
$$\simeq 14.2 \text{ weeks.}$$

When the growth rate is 20%,

$$\text{doubling time} = \frac{\log_{10} 2}{\log_{10} 1.2}$$
$$\simeq 3.8 \text{ weeks.}$$

Notice that multiplying the growth rate by 4 does not result in the doubling time becoming exactly a quarter of what it was.

Activity 40

From formula (14), the time taken for the RPI to double when the rate of inflation is 2% is

$$\frac{\log_{10} 2}{\log_{10} 1.02} = 35 \text{ years (to the nearest year).}$$

Similarly, the time taken for the RPI to double when the rate of inflation is 25% is

$$\frac{\log_{10} 2}{\log_{10} 1.25} = 3.1 \text{ years (to the nearest 0.1 year).}$$

Activity 41

Take formula (14)

$$d = \frac{\log_{10} 2}{\log_{10} b}$$

and multiply both sides by $\log_{10} b$ to get

$$d \times \log_{10} b = \log_{10} 2.$$

Then divide by d to obtain

$$\log_{10} b = \frac{\log_{10} 2}{d}.$$

So

$$b = 10^{(\log_{10} 2)/d} = 2^{1/d} = \sqrt[d]{2}.$$

Activity 42

Let x be the age of the sample in years. Then

$$(0.999\,879)^x = 0.71.$$

Hence

$$x = \frac{\log_{10} 0.71}{\log_{10} 0.999\,879}$$
$$\simeq 2830.$$

So the age of the sample is nearly 3000 years, and the date of the site can be estimated at around 1000 BCE.

Activity 43

The assumption is that the ratio of C^{14} to C^{12} in the atmosphere was the same when the carbon was fixed in the sample as it is now. This assumption turns out not to be entirely accurate in practice (largely because of the high quantity of fossil fuels burned in recent years). Consequently, the model should be modified to take account of this.

Activity 44

(i)   For a monthly interest rate of 1%, the APR can be calculated from

$$(1.01)^{12} = 1.1268\ldots.$$

This gives an APR of 12.68%.

(ii)   If the APR is 12% and the corresponding monthly interest rate is $x\%$, then

$$\left(1 + \frac{x}{100}\right)^{12} = 1.12$$
$$1 + \frac{x}{100} = \sqrt[12]{1.12}$$
$$\frac{x}{100} = \sqrt[12]{1.12} - 1,$$

hence

$$x = 100 \left(\sqrt[12]{1.12} - 1 \right)$$

$$= 0.95 \text{ (to 2 d.p.).}$$

So the monthly interest rate is 0.95% (to 2 d.p.).

Activity 45

The APR corresponding to a daily interest rate of 0.07% is $(1.0007)^{365} \times 100 - 100$, which is 29.10% (to 2 d.p.).

Activity 46

(a) The APR is 6.75%.

(b) The APR is $(1.0325)^2 \times 100 - 100 = 6.61\%$ (to 2 d.p.).

(c) The APR is $(1.005)^{12} \times 100 - 100 = 6.17\%$ (to 2 d.p.).

So option (c) results in you paying the least interest overall.

Activity 47

Again, calculate the APR, but this time choose the option with the higher APR.

(a) The APR is
$(1.0003)^{365} \times 100 - 100 = 11.57\%.$

(b) The APR is $(1.002)^{52} \times 100 - 100 = 10.95\%.$

The daily rate in option (a) is better for a saver.

Activity 48

Take one example: the calculation for compounding the interest 52 times per year, which is as follows.

With an APR of 100%, the interest on a loan of £1 after 1 year would be £1. Dividing this by 52 gives a notional interest rate of $\frac{1}{52}$. The actual amount owed based on this rate would be given by $(1 + \frac{1}{52})^{52} = 2.692\ldots$. So, rounding to the nearest penny, the total sum owing at the end of one year would be £2.69.

The completed table is as follows.

Number of times per year interest is compounded	Total sum owing at end of one year /£
1	2
2	2.25
4	2.44
12	2.61
24	2.66
52	2.69
365	2.71
500	2.7156
1000	2.7169
10000	2.7181

Activity 49

(a) The scatterplot should look like this:

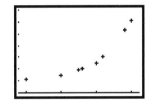

(b) Using exponential regression, the best fit is

$$y = 0.309 \times 1.012^x,$$

where y is the world population and x is in years. Therefore, $a = 0.309$ and $b = 1.012$.

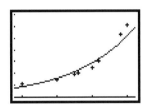

You should see that the exponential model slightly overestimates the actual data between 1925 and 1960, but underestimates the data for 1990 and 2000.

Activity 50

In Activity 49, the growth factor was found to be 1.012.

Using formula (14) gives the doubling time, d, as

$$d = \frac{\log_{10} 2}{\log_{10} 1.012}$$

$$= 58 \text{ years (to the nearest year).}$$

So the doubling time is about 58 years, based on taking the rounded value $b = 1.012$. Using the unrounded value $b = 1.011843507$ gives a doubling time of 59 years (to the nearest year).